Conference Proceedings of the Society for Experimental Mechanics Series

Series Editor

Kristin B. Zimmerman, Ph.D.
Society for Experimental Mechanics, Inc.,
Bethel, CT, USA

More information about this series at http://www.springer.com/series/8922

Pablo Zavattieri • Majid Minary • Martha Grady • Kathryn Dannemann
Wendy Crone
Editors

Mechanics of Biological Systems, Materials and other topics in Experimental and Applied Mechanics, Volume 4

Proceedings of the 2017 Annual Conference on Experimental and Applied Mechanics

Springer

Editors
Pablo Zavattieri
Purdue University
West Lafayette
IN, USA

Martha Grady
College of Engineering
University of Kentucky
Lexington, KY, USA

Wendy Crone
Department of Engineering Physics
University of Wisconsin – Madison
Madison, WI, USA

Majid Minary
Department of Mechanical Engineering
University of Texas at Dallas
Richardson, TX, USA

Kathryn Dannemann
Rensselaer Polytechnic Institute
Troy, NY, USA

ISSN 2191-5644 ISSN 2191-5652 (electronic)
Conference Proceedings of the Society for Experimental Mechanics Series
ISBN 978-3-319-63551-4 ISBN 978-3-319-63552-1 (eBook)
DOI 10.1007/978-3-319-63552-1

Library of Congress Control Number: 2015935194

Printed on acid-free paper

This Springer imprint is published by Springer Nature
The registered company is Springer International Publishing AG
The registered company address is: Gewerbestrasse 11, 6330 Cham, Switzerland

Preface

Mechanics of Biological Systems, Materials and Other Topics in Experimental and Applied Mechanics represents one of nine volumes of technical papers presented at the 2017 SEM Annual Conference and Exposition on Experimental and Applied Mechanics organized by the Society for Experimental Mechanics and held in Indianapolis, IN, June 12–15, 2017. The complete proceedings also includes volumes on *Dynamic Behavior of Materials; Challenges in Mechanics of Time-Dependent Materials; Advancement of Optical Methods in Experimental Mechanics; Micro- and Nanomechanics; Mechanics of Biological Systems and Materials; Mechanics of Composite, Hybrid and Multifunctional Materials; Fracture, Fatigue, Failure and Damage Evolution; Residual Stress, Thermomechanics and Infrared Imaging, Hybrid Techniques and Inverse Problems; and Mechanics of Additive and Advanced Manufacturing.*

Each collection presents early findings from experimental and computational investigations on an important area within experimental mechanics, the mechanics of biological systems and materials, and other topics in experimental and applied mechanics such as education and research in progress, to name a few.

The biological systems and materials segment of this volume summarizes the exchange of ideas and information among scientists and engineers involved in the research and analysis of how mechanical loads interact with the structure, properties, and function of living organisms and their tissues. The scope includes experimental, imaging, numerical, and mathematical techniques and tools spanning various lengths and time scales. Establishing this symposium at the Annual Meeting of the Society for Experimental Mechanics provides a venue where state-of-the-art experimental methods can be leveraged in the study of biological and bio-inspired materials, traumatic brain injury, cell mechanics, and biomechanics in general. A major goal of the symposium was for participants to collaborate in the asking of fundamental questions and the development of new techniques to address bio-inspired problems in society, human health, and the natural world. The organizers would like to thank all the speakers and staff at SEM for enabling a successful program.

West Lafayette, IN, USA — Pablo Zavattieri
Richardson, TX, USA — Majid Minary
Lexington, KY, USA — Martha Grady
Troy, NY, USA — Kathryn Dannemann
Madison, WI, USA — Wendy Crone

Contents

Chapter 1
Design of Bolted Connection in Composite Beams for Moment Resistance

H. K. Cho, J. M. Considine, D. R. Rammer, and R. E. Rowlands

Abstract Bolted/pinned joints in orthotropic composite materials have received considerable attention over the years. Bolt fastening is one of the most commonly used methods to connect wood to wood and/or wood to steel, etc. Stresses at such connections can be the "Achilles' heel", causing structural failures. Notwithstanding the challenges in stress analyzing bolted joints, their advantages and widespread use motivate developing ability to optimize their design.

Acknowledging the above, a finite element code is combined here with a screening optimization algorithm to optimize a bolt-hole pattern used to connect orthotropic wood members. A loaded wood beam having four connecting bolt holes at one end is optimized. The ultimate goal is to find optimal hole pattern and/or individual hole position under given load and displacement boundary conditions.

Keywords Optimization • Orthotropic material • Wood • Hole • Bolted joints • FEA

1.1 Introduction

Motivated by features such as ease of assembly, bolted joints are commonly used in steel and wood structures. Although considerable related literature exists on the mechanics of loaded holes [1–4], little appears to be available on optimizing bolted connections in wood. Wood is a natural and recyclable orthotropic material which is receiving extensive current attention for multi-story structures. Beam connections are critical for all multi-story wood structures, but especially in earthquake prone zones.

Optimization of bolt-hole pattern is carried out for a loaded wooden (Douglas Fir) beam so as to minimize the bearing stresses at the holes and beam deflection. Four bolt holes are involved. While theoretical formulae are available for analyzing stresses and strains around loaded holes, such studies tend to be limited to a single hole in an infinite member. In order to reduce the number of experiments examining bolt hole configuration, a FEM analysis was the primary component of the initial investigation.

An optimization algorithm is combined with FEA module to optimize the design. Commercial FEA software, ANSYS Xplorer, is used for the optimization. The main solver consists of a general static analysis FEA code and optimization algorithm. The process iterates until convergence of the optimization algorithm is reached. Since numerous iterations are needed to obtain an accurate solution, the iteration process time-consuming. During the process, the geometric model configuration changes at every step. Corresponding to the model shape modifications, new mesh-generation, application of boundary condition and static stress analysis are conducted.

Since the maximum contact stress at hole boundaries are a criterion for the optimization analysis, accurate values of these bearing stresses are necessary, though this effort examines the 2-dimensional case only. Several numerical and/or experimental studies appear in the literature which attempt to account for the actual bolt-hole contact stresses. In the present

H.K. Cho (✉)
Exto Engineering Co. Ltd, Hayangup Hayangro 13-13, Kyungsan Kyungpook, 37430, South Korea
e-mail: marklee1@hanmail.net

J.M. Considine • D.R. Rammer
USDA Forest Products Laboratory, Madison, WI, 53705, USA

R.E. Rowlands
University of Wisconsin, Madison, WI, 53706, USA

P. Zavattieri et al. (eds.), *Mechanics of Biological Systems, Materials and other topics in Experimental and Applied Mechanics, Volume 4*, Conference Proceedings of the Society for Experimental Mechanics Series, DOI 10.1007/978-3-319-63552-1_1

study, bolt contact conditions with a special function is adopted which well represents the contact bolt-hole phenomena. The iteration software process including both FEA and optimization algorithm enables one to find several optimal hole positions in the elastic wood plate.

1.2 Optimization Architecture

Optimization is achieved with ANSYS Xplorer. The method performs the theoretical background and processes the optimization according to a slightly different technique than most conventional optimization numerical methods. The conventional methods search the highest and lowest points by calculating the slope of the design domain, or random search method like genetic algorithm which generates a number of candidates and repeat iteration to reach the ultimate point. The method is called 'natural selection and degeneration.

The present optimization process is shown in Fig. 1.1. The individual optimization and FEA module mutually exchange results to obtain a solution. The optimization module defines the design variables and then determines the sampling points by the well-known DOE (design of experiment) method [5]. It provides a screening set to determine the overall trends of the meta-model to better guide the choice of options in optimal space filling design.

The finite element analysis results at individual sampling points are calculated to obtain an exact solution at a specific point. The ultimate goal of design sampling point (design point) determination by such an experimental design method is to find the "response surface" in the design area to be used for the final optimization. Constructing the response surface as

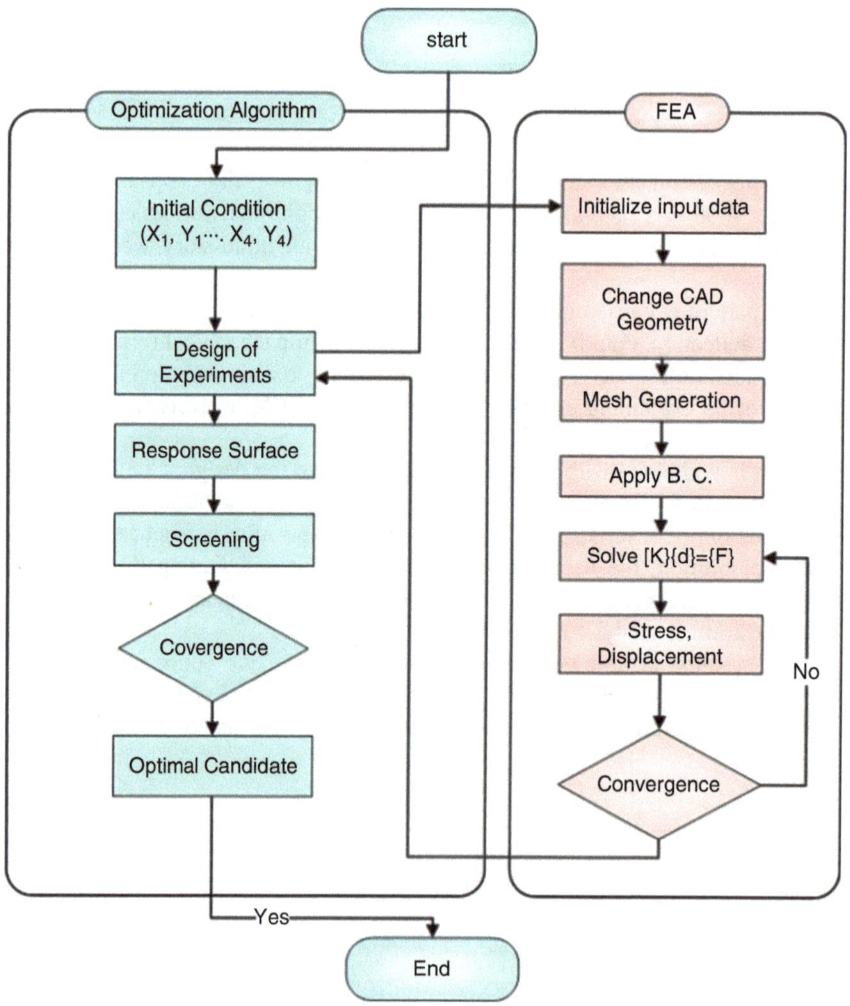

Fig. 1.1 Schematic diagram of the optimization architecture process

close to the actual solution as possible is important in determining the accuracy of the analysis. The accuracy of a response surface depends on several factors: complexity of the variations of the solution, number of points in the original design of experiments and the choice of the response surface type [5].

Various theories have been utilized to calculate the response surface. A Kriging scheme is used here to better represent complex nonlinear design surfaces. After the response surfaces have been computed, the design can be thoroughly investigated using a variety of numerical tools and valid design points identified by optimization techniques. We used the screening methods as an optimization method to obtain the final optimal solution. The screening scheme, which can be used for response surface optimization, allows one to generate a new sample set and sort its samples based on objectives and constraints. It is a non-iterative approach that is available for all types of input parameters.

The method effectively distributes a large number of candidates in the entire design domain and then provides the best several results through an accurate assessment. This optimization process is called "goal driven optimization". Advantages of this method are that the optimal solution is very effective in preventing entrapped local minimum and/or maximum, and it easily and rapidly finds a global optimal point. The computation time is also much less than that of the genetic algorithm approach which finds the optimal solution by the conventional random method. A disadvantage of the method is that the numerical calculation processes are relatively complex and the accuracy of the optimal solution may be somewhat reduced if the response surface is not correctly constructed.

1.3 Application

1.3.1 Material Properties of Wood Beam

An optimization analysis has been carried out here on a wood (Douglas-Fir) beam containing bolt holes. Wood is n orthotropic composite material and its mechanical properties are well known, Fig. 1.2 [6]. Relative to bolted joints, the present objective is to enhance the mechanical performance through the optimization of the bolt hole. The member used in this analysis is a relatively thin wooden beam, and due to a negligibly small variation of the material properties through the thickness, the wood beam can be implemented as a 2D (plane stress) model.

Table 1.1 shows the properties of Douglas-Fir wood [6, 7] need for the FEM analysis. The stiffness and physical properties of wood have some different characteristics from the physical properties of metal.

The effect of temperature on material properties is relatively small in wood, but the effect of moisture content is relatively large. For materials with very high longitudinal and transverse stiffness, such as wood, much attention is needed in stress analysis, e.g., stress distributions can be highly influenced by the material directionality. This can be particularly important

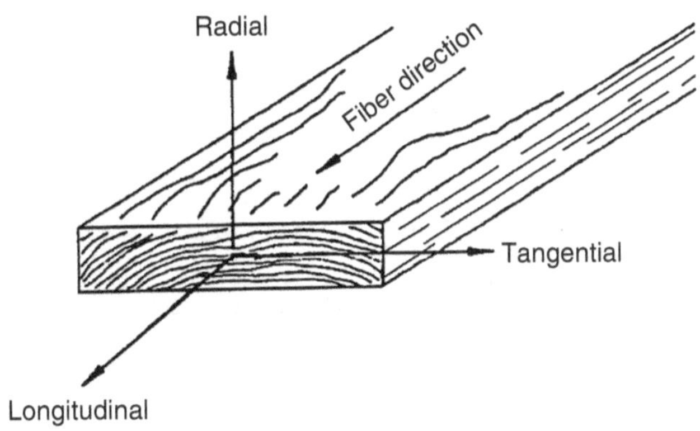

Fig. 1.2 Orthotropic material directions in wood

Table 1.1 Material properties of Douglas-Fir

Properties	
E_L (GPa)	13.53
E_T (GPa)	0.77
G_{LT} (GPa)	1.06
ν_{LT}	0.45

at the stress concentrations near bolt holes. In addition, unlike ordinary metal materials, wood exhibits higher longitudinal tensile than compressive strengths. Table 1.2 shows strengths of Douglas-Fir.

1.3.2 Optimization Features

The wood beam analyzed is 305 cm × 56 cm with four bolt holes on the left end for connection. Beam thickness is 25.4 mm and the diameter of the bolt holes is 19.1 mm (3/4 in.). The boundary conditions are shown in Fig. 1.3. As shown in Fig. 1.3, a symmetric boundary condition is applied at the right end, and a 0.5 MPa uniformly distributed load is applied along the top face of the beam.

The problem definition of optimization of the bolt hole position is as follows. First, the design area for the holes shown in Fig. 1.4 is a rectangular area of 51 cm × 56 cm at the left end of the beam. The goal is to determine the optimum location within this area of the four individual holes upon loading the beam. The objective function for optimization is defined as a multi-objective function which combines the vertical deflection at the left end of the rectangular wood beam in Fig. 1.3, **a**, and maximum von Mises stress around the holes. The optimization process minimizes the objective function. A symmetric boundary condition is imposed at the right end of the beam.

Eight coordinates (x_1, y_1, x_2, y_2, x_3, y_3, x_4, y_4), two for each hole, were defined as the design parameters. As shown in Fig. 1.3, the *x*-coordinate represents the straight line distance measured from the left end to the hole center, and the y-coordinate is defined as the absolute value of the straight line distance from the upper and lower sides of the beam to the hole center.

Objective function

$$F\left(x_1, y_1, \cdots, x_4, y_4\right) = \delta_v(a) + \sigma_{\max .von\ Mises}\ (around\ hole) \tag{1.1}$$

Design variables.

$$x_1, y_1, x_2, y_2, x_3, y_3, x_4, y_4 \tag{1.2}$$

Table 1.2 Directional strengths of Douglas-Fir

Strength parameter	
Longitudinal tensile (GPa)	90.0
Longitudinal compression (GPa)	47.6
Tangential tensile (GPa)	2.7
Tangential compression (GPa)	5.3
Shear (longitudinal – Tangential) (GPa)	9.7

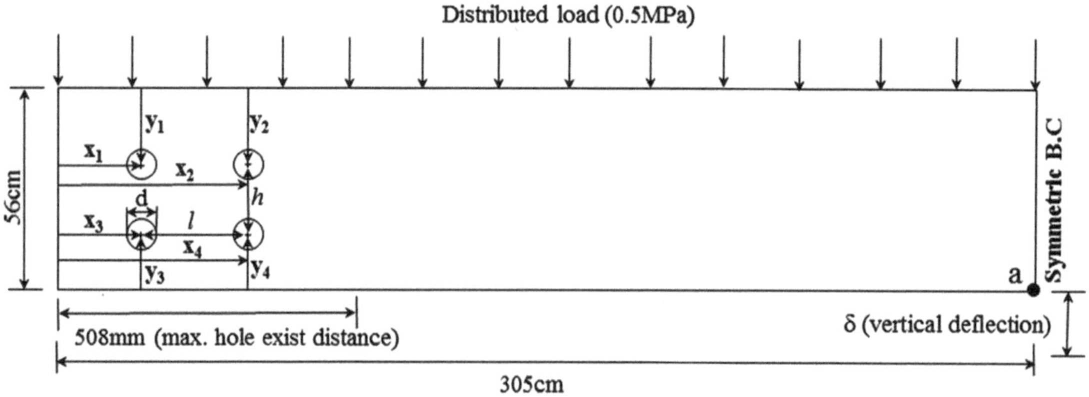

Fig. 1.3 Geometry configuration and dimension of the optimization analysis problem

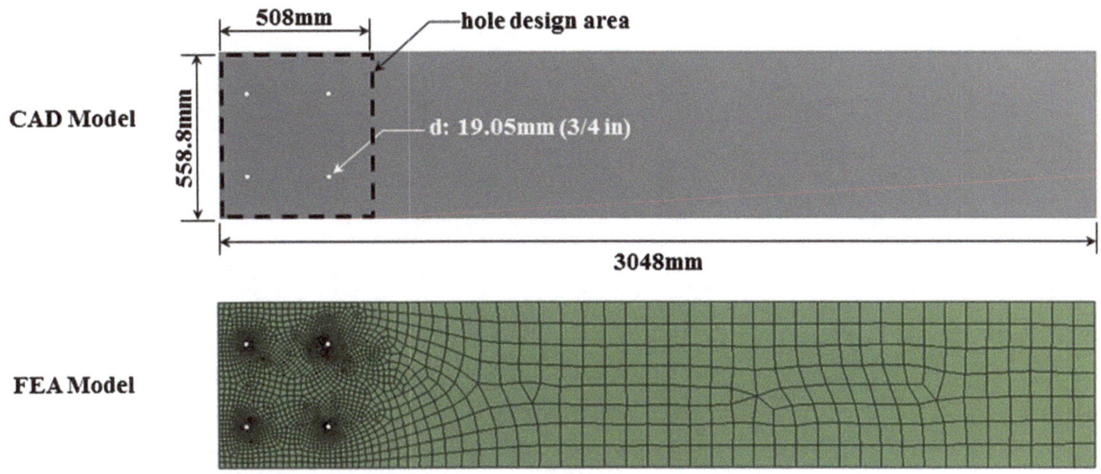

Fig. 1.4 CAD geometry and example FEA model employed. FE mesh changes with each analysis

Constraints.

$$\begin{aligned} &1)\ 4d \le x_1, x_3 \le 7d \\ &2)\ y_1, y_2, y_3, y_4 \ge 2d \\ &3)\ 1.5d \le \mathrm{h} \le 16d \\ &4)\ \mathrm{l} \ge 4d \end{aligned} \tag{1.3}$$

where d = 19.1 mm is the bolt diameter.

The constraints applied to the optimization analysis are shown in Eq. 1.3 and represent position conditions which should be satisfied by each hole after the optimization process. The positions of holes 1 and 3 in the x direction should be greater than $4d$ and less than $7d$. The geometric constraints also define the distance between the holes and the minimum distance from the outer edges of the rectangular beam. The optimization proceeds by finding the response surface and then randomly distributing a number of candidates on the response surface to obtain an optimal solution. Many analyses are therefore needed to accurately construct such response surfaces in the optimization calculations. Upon each analysis step, both the positions of the holes and the shape of the FEM mesh changes.

The FE model shown in Fig. 1.4 represents a mesh using the ANSYS automatic of mesh generation. One of the most important results of the FEA model is determination of the stress concentrations at each hole. This phenomenon is very significant in orthotropic materials, such as wood, which have different moduli in the longitudinal and transverse directions. The sizes of the elements immediately adjacent to each hole are small (less than16 mm^2) in order to enhance the accuracy of stress calculation. The area of the element is about 1.4% of the hole area. The entire model was divided into finite elements by the auto mesh process. The aspect ratio is set to be smaller than 2, and Jacobian is set to be 0.3 or more. Since the shape of the model is not complicated, the mesh generation process converges well.

The right section of the plate is meshed with the relatively large elements because there are no steep stress gradients. Compatible with prevalent practice, the present boundary conditions correspond to those for rigid bolts, and zero bolt-wood friction and bolt clearance. The contact response between the hole and the bolt is automatically determined and calculated based on these conditions.

1.4 Results

The deflection at the right end of the beam, and the von Mises stress at each hole, were simultaneously determined as the objective function and optimization is performed to minimize the end deflection.

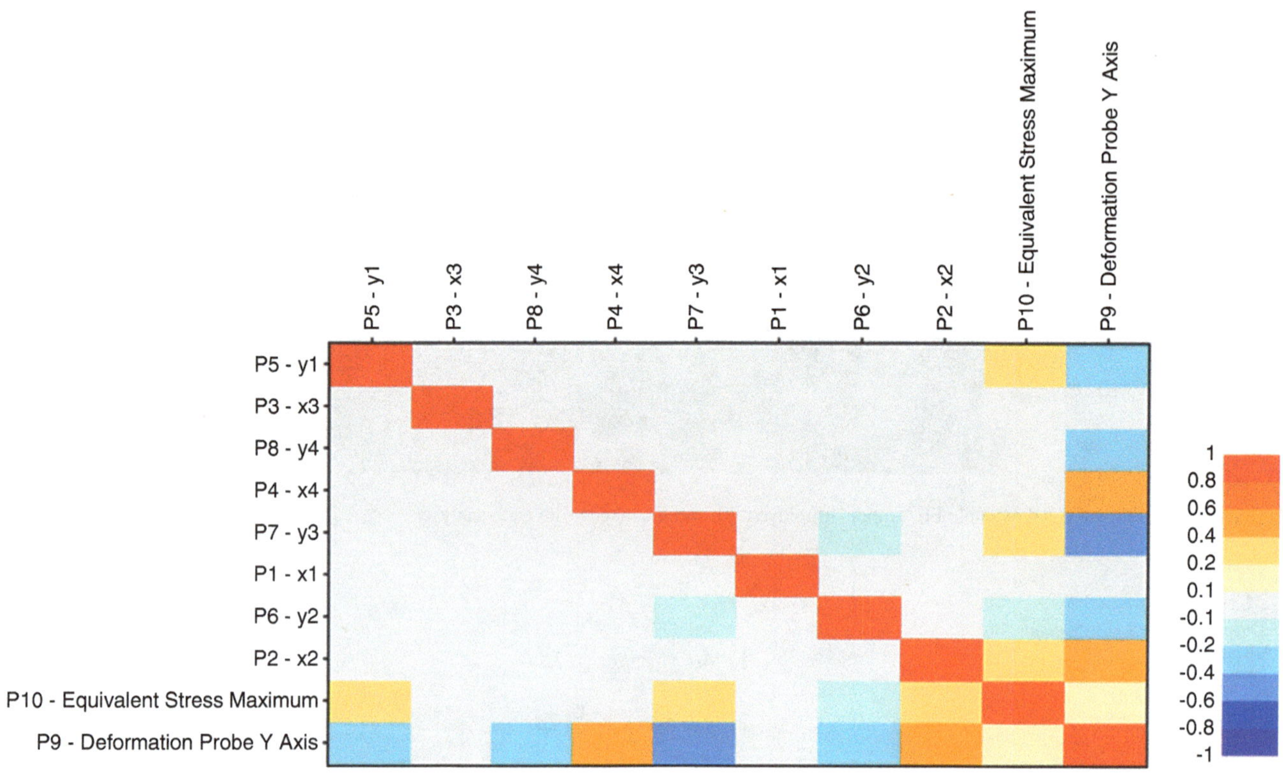

Fig. 1.5 Correlation matrix of the design variables and objective function

Table 1.3 Optimization results

	Hole location								Results			
	Hole 1		Hole 2		Hole 3		Hole 4					
Name.	x_1(mm)	y_1 (mm)	x_2 (mm)	y_2 (mm)	x_3 (mm)	y_3 (mm)	x_4 (mm)	y_4 (mm)	Max. von Mises stress (MPa)	Enhance -ment (%)	Vertical deflection (mm)	Enhance -ment (%)
Initial	95.3	139.5	381.0	139.5	95.3	139.5	381.0	139.5	**147.5**	**0**	**13.0**	**0**
Candidate 1	119.5	74.8	348.8	158.5	91.6	81.7	324.6	40.1	**125.7**	**14.8**	**11.9**	**9.1**
Candidate 2	132.0	87.7	472.7	233.5	125.1	119.2	480.1	70.8	**126.6**	**14.1**	**10.9**	**16.6**
Candidate 3	111.3	114.1	420.8	245.5	78.6	69.8	262.4	72.9	**116.4**	**21.1**	**12.9**	**1.0**

As a preliminary step towards optimization, the strength of dependency between design variables and objective functions is calculated and presented in the form of a matrix in Fig. 1.5. The closer to −1 and 1 in Fig. 1.5, the higher the interdependence. It can be seen that the deflection is more dependent on the positional parameters of the hole than the stress, and the position of hole 4 and hole 2 has a great influence on point **a** deflection in Fig.1.3. Interpretation of these charts can improve understanding of overall design and infer design criteria.

Three 'best' bolt-hole configurations were suggested through optimization analysis and are shown in Table 1.2. Results of deflection and stress for each of the three solutions (Candidates 1, 2 and 3) are compared with those of the initial design. In the initial design, the von Mises stress was 147.5 MPa, but for optimized candidates 1, 2 and 3, the stresses are 125.7 MPa, 126.6 MPa, and 116.4 MPa, respectively, which decreased by 14.8%, 14.1%, and 21.1%, respectively. In the case of the end-beam displacement, the initial displacement was 13 mm, but that was decreased by 9.1%, 16.6% and 1.0%, respectively, under optimization. Among the candidates of this optimum design solution, the best result situation can be selected according to the design purpose such as consideration of displacement and/or stress.

Figure 1.6 shows the hole locations for the three optimization candidates. The positions of the initial holes are represented by circles, and the optimized hole patterns are represented by filled black circles. The position of each relocated hole satisfies all of the constraints given in Eq. 1.3, and the exact coordinates are shown in Table 1.3.

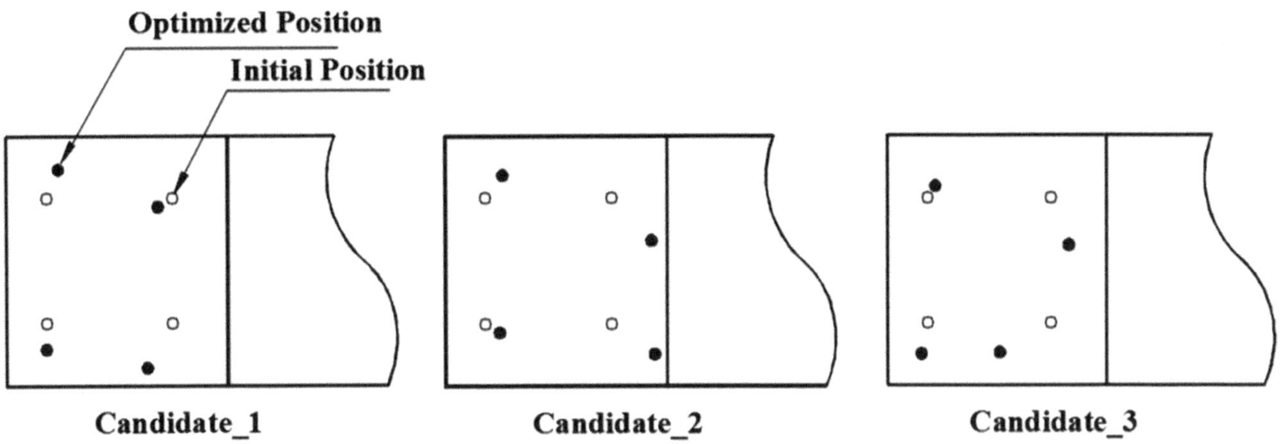

Fig. 1.6 Optimization results (three optimal hole-pattern candidates)

1.5 Summary, Discussion and Conclusions

The procedure, methodology and the analyses on the optimization of the hole pattern for a bolted wood joint are presented. Optimization was performed on a wood beam subjected to a distributed load and containing four bolt holes near one end. Optimal hole positions were determined numerically to minimize the maximum bearing stress at the holes and the vertical deflection at the end of the hole and thereby providing increased moment resistance at the connection.

The deflection and stresses of the optimized hole pattern were 14 $\sim$ 20% more effective than the design of the original conventionally-located square pattern for the holes. When solving the optimization problem of determining the positions of multiple holes as here, the probability of finding a local maximum point and a minimum point by a combination of a plurality of hole patterns is very high.

There are many difficulties in solving the complex problem of determining the best positions of many holes such as using the conventional gradient optimization method. The optimization problem by the gradient method is very likely to find the local maximum and minimum points because the hole pattern optimization problem has a relatively flat design surface, providing many minima. Therefore, as a method for locating the global optimal point in a complex design domain, it is effective to distribute a large number of candidate groups in the entire design domain to obtain the best result.

The analytical procedures adopted here are highly applicable and could be used with small additional effort for supplementary holes in the bolt pattern.

Although many papers have addressed problems of stress analysis, design, and optimization of holes, the authors are unaware of any previous publication which optimizes the bolt-hole pattern involving multiple holes in an orthotropic material. Future considerations include combining an optimization such as here with a strength criterion and experimental information, including optical stress analyses.

Acknowledgements This research was supported by INNOPOLIS Foundation project ACC-2015-DGI-00546 and the US Department of Agriculture for which the authors are very grateful.

References

1. Toussaint, E., Durif, S., Bouchair, A., Grediac, M.: Strain measurements and analyses around the bolt holes of structural steel plate connections using full-field measurements. Eng. Struct. **131**(15), 148–162 (2017)
2. Xiwu, X., Liangxin, S., Xuqi, F.: Stress concentration of finite composite laminates weakened by multiple elliptical holes. Int. J. Solids Struct. **32**(20), 3001–3014 (1995)
3. Soutis, C., Fleck, N.A., Curtis, P.T.: Hole-hole interaction in carbon fiber/epoxy laminates under uniaxial compression. Composites. **22**(1), 31–38 (1991)
4. Yeh, H.Y., Le, M.D.: Mutual influence about stress concentration of holes in composite plates. J. of Reinforced Plastics and Composites. **12**, 38–47 (1993)
5. ANSYS Users Manual
6. Kunesh, R.H.: Strength and elastic properties of wood. For. Prod. J. **18**(1), 65–72 (1968)
7. USDA, Wood Handbook, General Technical Report FPL-GTR-190, 5-2 $\sim$ 5-7

Chapter 2
Measurement of 3D Deformation Field of ECM Generated by Mesenchymal Stem Cell Using DVC Method

Yasuyuki Morita, Naoki Kawase, Takashi Yamauchi, and Yang Ju

Abstract Cells maintain homeostasis and express various functions by interacting mechanically with cell-adhesive matrix. As regarding cellular differentiation, it has been found that elasticity of the matrix determines differentiation lineage of mesenchymal stem cells (MSCs). The direct quantitative measurement of the mechanical interaction between MSC and matrix for differentiation, however, has not been performed. In the present work, the displacement field of the cell-adhesive matrix was observed quantitatively using digital volume correlation (DVC) method. In practice, the cellular traction stress was analyzed when the MSC differentiated into neuron or osteoblast on the soft or hard elastic matrix, respectively. Then, function of non-muscle myosin II (NMM II), which plays an important role in intracellular cytoskeletal dynamics, was investigated in cellular differentiation. As a result, the mechanical interaction between the cell and the matrix was dependent upon the elasticity of the matrix. Additionally, it has been shown that mechanical interaction between intracellular cytoskeleton and cell-adhesion matrix is indispensable for cellular differentiation.

Keywords Digital volume correlation (DVC) • Elastic modulus • Mesenchymal stem cell (MSC) • Differentiation • Extracellular matrix (ECM)

2.1 Introduction

MSC (Mesenchymal Stem Cell or Multipotent Stromal Cell) has self-renewal ability and plasticity. The MSCs can differentiate into wide variety of cells such as osteoblasts, chondrocytes, adipocytes, myocytes, cardiomyocytes, neurons etc [1–3]. They must have great promise with respect to stable supply of various mature cells in regenerative medicine since they are harvested relatively easily from bone marrow. Therefore, control technique of the stem cell differentiation is indispensable from the view point of steady cell supply. Mechanical stimulus [4–8] would be one of the prospective technologies for controlling differentiation because it is simple and easy although diverse technologies such as chemical factor [9], electrical and magnetic stimulus [10], and co-culture system [11, 12] have been developed. It has been becoming increasingly clearer that cell differentiation has a close relationship with mechanical microenvironment of extracellular matrix (ECM) around the cell [13, 14]. Engler et al. elucidated that only elasticity of cell-adhesive matrix was able to determine cellular differentiation lineage [15]. This knowledge is one of the crucial achievements to carry forward cell-differentiation controls due to mechanical microenvironment. However, there have been no reports to bring out in terms of mechanical interaction between cell and ECM with different matrix elasticity.

In the present work differentiation lineage of human bone marrow MSCs (hBMSCs) and three dimensional displacement field of cell-adhesive matrix produced by a hBMSC were evaluated quantitatively when the hBMSC was cultured on the two kinds of considerably different elasticity $E = 0.5$ kPa (soft) and 37 kPa (hard) in the cell-adhesive matrix. The differentiation assessment was performed by two ways morphological and gene expression evaluations. The 3D displacement field was analyzed by a digital volume correlation (DVC) method [16] using a confocal laser scanning microscope (CLSM) (FN1CSU; Nikon Instech). The 3D–traction stress of the cell was also obtained quantitatively. Furthermore the same evaluations the differentiation assessments and the gene expression levels were performed if intracellular non-muscle myosin II (NMM II) which plays an important role in cytoskeletal dynamics inside cell [17] was inhibited.

Y. Morita (✉) • N. Kawase • T. Yamauchi • Y. Ju
Department of Mechanical Science & Engineering, Graduate School of Engineering, Nagoya University, Furo-cho, Chikusa-ku, Nagoya, 464-8603, Japan
e-mail: morita@mech.nagoya-u.ac.jp

P. Zavattieri et al. (eds.), *Mechanics of Biological Systems, Materials and other topics in Experimental and Applied Mechanics, Volume 4*, Conference Proceedings of the Society for Experimental Mechanics Series, DOI 10.1007/978-3-319-63552-1_2

2.2 Materials and Methods

2.2.1 *Cell Culture*

HBMSCs (UE6E7T-3; Riken Cell Bank) were maintained in low-glucose Dulbecco's modified Eagle's medium (DMEM; Wako Pure Chemical Industries, Ltd.) containing 10% fetal calf serum (Invitrogen) and 0.05% gentamicin (Invitrogen) at 37 °C under an atmosphere of 5% CO_2 in a humidified incubator. At near-confluence, which occurred every 5–7 days, the cells were detached from 25 cm^2 culture flasks (BD Biosciences) with 0.25 w/v% Trypsin-1 mM EDTA (Wako Pure Chemical Industries, Ltd.). Actin-GFP (Molecular Probes) was added into the suspension of the hBMSCs to stain the intracellular actin filaments for cellular morphology observation. Then, the cells were seeded onto polyacrylamide gel whose surface covered with a cellular adhesion molecule, human fibronectin (FN; Corning Incorporated Life Sciences), as described in the following sections.

2.2.2 *Preparation of Polyacrylamide-Matrix Gel*

Solutions of acrylamide (10 w/v%; Kanto Chemical Co., Inc., Tokyo, Japan) and N, N-methylene-bis-acrylamide (BIS; Kanto, Chemical Co., Inc.) were mixed with distilled water to fabricate the polyacrylamide. The blending quantity of BIS was 0.003 and 0.07 w/v% to change elasticity of the polyacrylamide. For measurement of 3D-displacement field by DVC method, fluorescent micro-particles (1.0-μm diameter, Molecular Probes) suspension was added to the polyacrylamide solution with 10 w/v% ammonium persulfate (Kanto, Chemical Co., Inc.) and TEMED (Amresco) which work as polymerization initiator and accelerator, respectively, in a volume ratio of 100 (polyacrylamide): 9 (micro-particles): 1 (ammonium persulfate): 1 (TEMED). 6 μl of the acrylamide solution was pipetted on the surface of a glass based dish which coated with aminosilane (Iwaki). Then, a cover glass (10-mm diameter, Matsunami Glass) was put onto the solution to make the top surface flat. Polymerization was allowed to proceed at room temperature for 5 min. Then, the polyacrylamide-matrix gel was thoroughly rinsed with distilled water. The typical geometry of the matrix was approximately 10 and 80 μm in diameter and thickness, respectively. The two kinds of elastic moduli of the polyacrylamide gel contained BIS of 0.003 or 0.07 w/v%, were 0.5 or 37 kPa, respectively. The elastic modulus of the matrix was determined by the penetration method [18].

2.2.3 *Coating of the Polyacrylamide Matrix with Human Fibronectin*

The human fibronectin was conjugated to the top surface of the acrylamide matrix using hetero-bifunctional crosslinker, sulfo-SANPAH (Terumo), to promote cell attachment to the polyacrylamide matrix. The typical thickness of the fibronectin layer was approximately 1 μm as determined by CLSM.

2.2.4 *Cell Seeding and Differentiation Evaluation*

The hBMSCs were seeded onto the matrix with low density of 1.6×10^4 cells, and cultured for 24 h. Two kinds of evaluation techniques for cellular differentiation, morphological assessment and gene expression analysis, were employed. In the morphological assessment, parameters of branches B and roundness R were adopted for evaluation of neurogenic and osteogenic differentiation, respectively. Meanwhile, enolase 2 (Eno2) and runt-related transcription factor 2 (Runx2) gene expression levels were analyzed using quantitative reverse transcription polymerase chain reaction (qRT-PCR) for the another differentiation evaluation, respectively. Eno2, commonly included in axon processes of neurons, is one of the specific enzymes of glycolysis [19]. Runx2 is an essential transcription factor for osteogenic differentiation [20]. Figure 2.1 shows the experimental arrangement in this study.

2.2.5 *DVC Measurement and Calculation of Displacements, Strains, and Stresses*

The full field 3D displacements of the polyacrylamide matrix were determined using CLSM and DVC method [21]. The DVC algorithm was referred to Ref. 16. In brief, confocal volume images of cell-seeded polyacrylamide matrix with randomly dispersed fluorescent particles were obtained after 24 h which the hBMSCs were seeded onto the matrix. The acquired

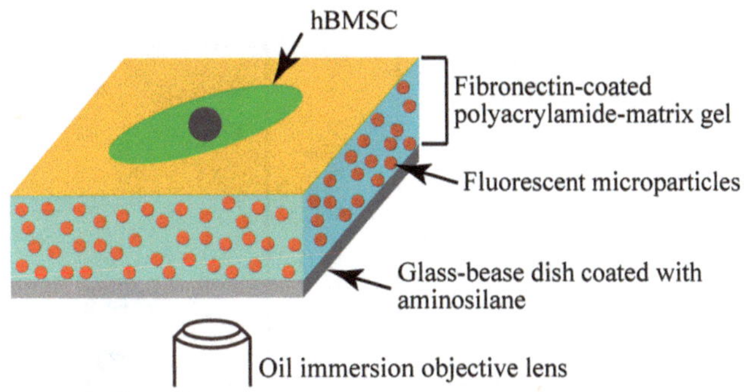

Fig. 2.1 Experimental arrangement. Fluorescent microparticles (1.0 μm diameter) were spread inside polyacrylamide-matrix gel for 3D displacement measurement by DVC. The typical geometry of the matrix was approximately 10 mm and 80 μm in diameter and thickness, respectively. Human fibronectin (approximately 1 μm thickness) was coated on the top surface of the matrix. HBMSCs were seeded onto the matrix at a density of 2.0×10^2 cells/mm^2. 3D confocal image stacks were acquired by CLSM that captured the emission from the fluorescent microparticles

volume of the matrix was $300 \times 300 \times 50$ μm^3, corresponding with $512 \times 512 \times 167$ voxels. Then, the confocal volume images were also obtained at the same position after 15 min. The stacked images were divided into a set of subvolumes that were centered on an area of interest. The local displacement vector can be obtained using a pair of corresponding subvolumes images through a three-dimensional volume correlation method. This was performed mathematically by representing the intensity patterns of each subvolume by two intensity functions $f(\boldsymbol{x})$ and $g(\boldsymbol{x})$ that have been translated relative to one another by a displacement vector $\boldsymbol{c}$ of each subset within the confocal volume image can be calculated through the cross-correlation of the two subset intensity functions [22]. The DVC algorithm programmed in this study was achieved displacement sensitivity of 41 nm with the CLMS. Finally, from the result of the 3D displacement field, traction stress of the cell was calculated along the top surface of the matrix, directly beneath the cell using the known Cauchy relationship: $\boldsymbol{T} = \sigma\, \boldsymbol{n}$, where $\boldsymbol{T}$ is the 3D surface traction stress vector, σ is the material stress tensor, and $\boldsymbol{n}$ is the surface normal. The complete mathematical details for determination of the traction stress can be found in Ref. 23.

2.2.6 NMM II Inhibition

As the final experiment, intracellular non-muscle myosin II (NMM II), which is likely to be involved in exerting force through focal adhesions in mechanisms of sensing matrix elasticity [15], was inhibited by blebbistantin (Sigma-Aldrich). The blebbistantin was added with the concentration of 50 μM after 6 h following the cells were seeded onto the polyacrylamide matrix.

2.2.7 Statistical Analysis

The number of samples, n, was indicated in each figure. The paired Student's t test was used for statistical analyses, and p values <0.01 and 0.05 were taken to indicate statistical significance.

2.3 Results and Discussion

The results of the cells without NMM II inhibitor whose data are plotted in red of Figs. 2.2 and 2.4, have been focused on initially. Figure 2.2a shows the morphological analysis for cellular differentiation with the elastic modulus of the polyacrylamide matrix. The cellular morphology was dependent upon the elastic modulus. There were also big differences with respect to the mRNA expression levels of Eno2 and Runx2 in the two kinds of elastic modulus of the matrix as shown in Fig. 2.2b. Namely, the maximum value of the expression levels for each marker, Eno2 and Runx2, was obtained on the matrix of $E = 0.5$ and 37 kPa, respectively. These results indicate that the differentiation of hBMSCs was dependent upon

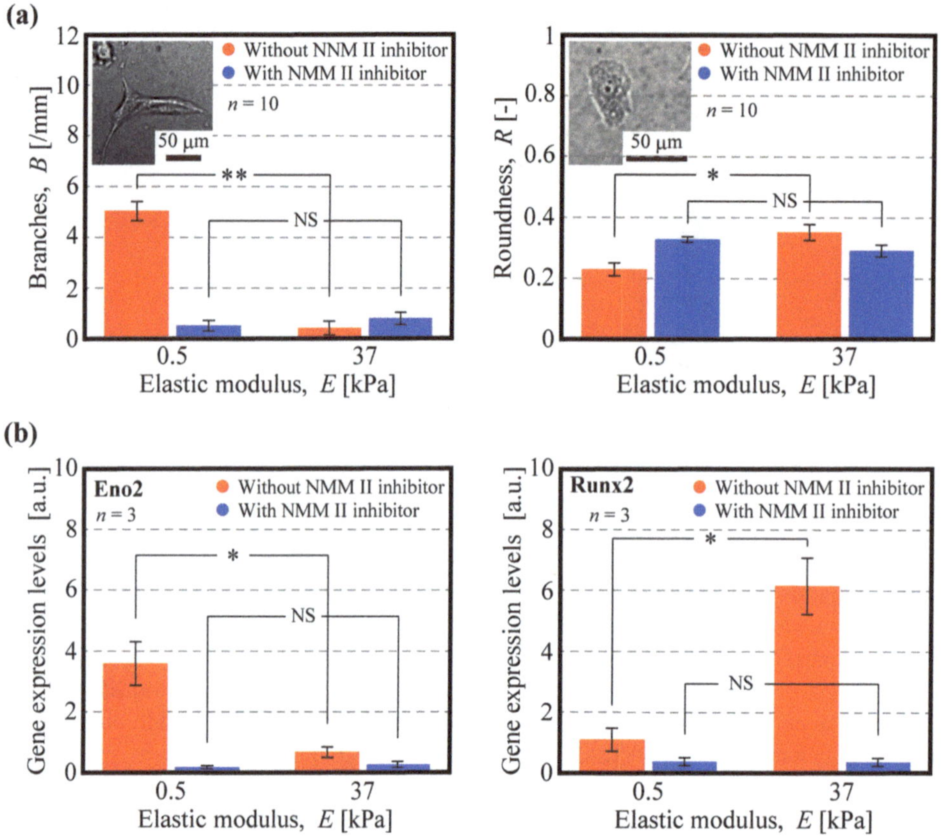

Fig. 2.2 Evaluation of differentiation of hBMSCs with elastic mudululs of polyacrylamide matrix. (**a**) Morphological analysis. Branches (*left*), roundness (*right*). *Inset* indicates the representative phase-contrast cell image at the maximum value without NMM II inhibitor. (**b**) mRNA expression levels. Eno2 for neurons (*left*), Runx2 for osteoblasts (*right*). *Red* and *blue* data represent without and with NMM II inhibitor, respectively. *n* indicates the number of samples. All data are indicated by Mean ± SEM. $^{**}p < 0.01$, $^{*}p < 0.05$. NS = not significant

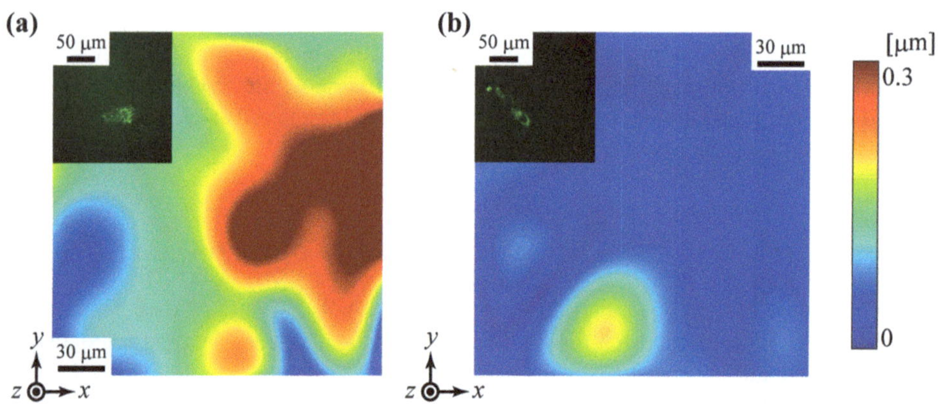

Fig. 2.3 Absolute displacement field of polyacrylamide matrix generated by cell without NMM II inhibitor. (**a**) Elastic modulus $E = 0.5$ kPa. (**b**) $E = 37$ kPa. Figures indicate the displacement field on the matrix. Inset indicates confocal fluorescence micrographs of the cell

the elasticity of the polyacrylamide matrix, and they underwent differentiation into neurons on $E = 0.5$ kPa, and osteoblasts on $E = 37$ kPa, respectively. This differentiation lineage agreed well with the findings of Engler et al. [15]. The displacement field generated by the cell on each elastic modulus of the matrix was shown in Fig. 2.3. The displacement denotes the absolute displacement. It suggests that a hBMSC produced large deformation (high displacement and wide area) to the matrix and differentiated into a neuron if the matrix was relatively soft. In contrast, it indicates that a hBMSC made small deformation (low displacement and narrow area) to the gel substrate and differentiated into a osteoblasts if the matrix was relatively hard.

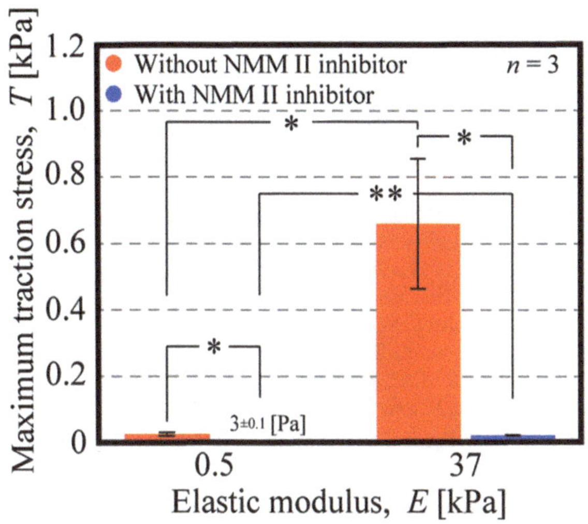

Fig. 2.4 Maximum traction stress, ***T***, of hBMSCs in each elastic modulus of the matrix. *Blue* and *red* data represent with and without the NMM II inhibitor, respectively. *n* indicates number of samples. All data are expressed as the means ± SEM. ** $p < 0.01$, * $p < 0.05$

Figure 2.4 shows the maximum cell-induced traction stress in each elastic modulus of the matrix. The maximum traction stress of the cell was dependent upon the elastic modulus of the matrix, and inversely correlated with the displacement trends shown in Fig. 2.3.

The maximum cell-induced traction stress of the respective elastic modulus is shown in blue in Fig. 2.4, when the intracellular NMM II is inhibited. There was big difference in terms of the traction stress values compared to it without NMM II inhibition (shown in red, Fig. 2.4). These results weren't attributable to cytotoxicity of NMM II inhibitor (blebbistantin). The results of the differentiation assessment were shown in Fig. 2.1. There were no clear differences with respect to the morphological evaluation and the gene expression levels in each elastic modulus of the matrix. It says that the cells didn't differentiate into either neurons or osteoblasts. NMM II is one of the vital factors of actin structures in intracellular cytoskeletal dynamics [24]. Furthermore, these actin structures are connected to focal adhesions which provide the pathway of force transmission from inside the cell to elastic matrix [25]. Therefore, it indicates that cytoskeleton plays an important role in mechanical interaction with elastic matrix [26], and acts a key function in cellular differentiation.

2.4 Conclusions

In summary, mechanical fields of elastic matrix and hBMSCs were measured using DVC method to investigate whether the elastic modulus of the matrix determined the differentiation lineage of hBMSCs. The hBMSCs produced large deformation to the matrix when the matrix was soft ($E = 0.5$ kPa) and opposite when the matrix was hard ($E = 37$ kPa). The traction stress of the cell produced, however, became higher when the matrix was hard. Moreover, it suggests that mechanical interaction between elastic matrix and intracellular cytoskeleton is crucial for cellular differentiation. Evaluating the mechanical fields in the mechanical interaction of stem cell differentiation has no precedent and will provide important knowledge in mechanobilogy of differentiation.

References

1. Pittenger MF, Mackay AM, Beck SC, Jaiswal RK, Douglas R, Mosca JD, Moorman MA, Simonetti DW, Craig S, and Marshak DR, Science 284, 143–147, (1999)
2. Dominici, M., Le Blanc, K., Mueller, I., Slaper-Cortenbach, I., Marini, F.C., Krause, D.S., Deans, R.J., Keating, A., Prockop, D.J., Horwitz, E.M.: Cytotherapy. **8**, 315–317 (2006)
3. Lazennec, G., Jorgensen, C.: Stem Cells. **26**, 1387–1394 (2008)
4. Morita, Y., Suzuki, S., Ju, Y., Kawase, N.: J Mech Med Biol. **14**, 1450025 (2014)
5. Morita, Y., Watanabe, S., Ju, Y., Xu, B.: Acta Bioeng. Biomech. **15**, 71–79 (2013)
6. Morita, Y., Mukai, T., Ju, Y., Watanabe, S.: Cell Biochem. Biophys. **66**, 73–80 (2013)
7. David, V., Martin, A., Lafage-Proust, H.M., Malaval, L., Peyroche, S., Jones, D.B., Vico, L., Guignandon, A.: Endocrinology. **148**, 2553–2562 (2007)
8. Thorpe, S.D., Buckley, C.T., Vinardell, T., O'Brien, F.J., Campbell, V.A., Kelly, D.J.: Ann. Biomed. Eng. **38**, 2896–2909 (2010)
9. Kratchmarova, I., Blagoev, B., Haack-Sorensen, M., Kassem, M., Mann, M.: Science. **308**, 1472–1477 (2005)

10. Zablotskii, V., Lunov, O., Novotna, B., Churpita, O., Trosan, P., Holan, V., Sykova, E., Dejneka, A., Kubinova, S.: Appl. Phys. Lett. **105**, 103702 (2014)
11. Fan, H.B., Liu, H.F., Toh, S.L., JCH, G.: Biomaterials. **29**, 1017–1027 (2008)
12. Morita, Y., Yamamoto, S., Ju, Y.: Biotechnol. Lett. **37**, 1911–1918 (2015)
13. Khetan, S., Guvendiren, M., Legant, W.R., Cohen, D.M., Chen, C.S., Burdick, J.A.: Nat. Mater. **12**, 458–465 (2013)
14. Chowdhury, F., Na, S., Li, D., Poh, Y.C., Tanaka, T.S., Wang, F., Wang, N.: Nat. Mater. **9**, 82–88 (2010)
15. Engler, A.J., Sen, S., Sweeney, H.L., Discher, D.E.: Cell. **126**, 677–689 (2006)
16. Franck, C., Hong, S., Maskarinec, S.A., Tirrell, D.A., Ravichandran, G.: Exp. Mech. **47**, 427–438 (2007)
17. Vicente-Manzanares, M., Ma, X.F., Adelstein, R.S., Horwitz, A.R.: Nat Rev Mol Cell Bio. **10**, 778–790 (2009)
18. Landau, L.D., Lifshitz, E.M.: Theory of Elasticity. 3rd ed. Sykes, J.B., Reid, W.H. (translators). Pergamon Press, Oxford (1986)
19. Clegg, N., Ferguson, C., True, L.D., Arnold, H., Moorman, A., Quinn, J.E., Vessella, R.L., Nelson, P.S.: Prostate. **55**, 55–64 (2003)
20. Brubaker, K.D., Vessella, R.L., Brown, L.G., Corey, E.: Prostate. **56**, 13–22 (2003)
21. Bay, B.K., Smith, T.S., Fyhrie, D.P., Saad, M.: Exp. Mech. **39**, 217–226 (1999)
22. Maskarinec, S.A., Franck, C., Tirrell, D.A., Ravichandran, G.: P Natl Acad Sci USA. **106**, 22108–22113 (2009)
23. Franck, C., Maskarinec, S.A., Tirrell, D.A., Ravichandran, G.: PLoS One. **6**, e17833 (2011)
24. McBeath, R., Pirone, D.M., Nelson, C.M., Bhadriraju, K., Chen, C.S.: Dev. Cell. **6**, 483–495 (2004)
25. Tamada, M., Sheetz, M.P., Sawada, Y.: Dev. Cell. **6**, 709–718 (2004)
26. Mathieu, P.S., Loboa, E.G.: Tissue Eng B. **18**, 436–444 (2012)

Chapter 3
Global Vibrational Modes in Proteins: Raman Spectroscopy and Numerical Modeling

Alberto Carpinteri, Giuseppe Lacidogna, Andrea Bassani, and Gianfranco Piana

Abstract Protein biological functions are made possible by folding, a quick change in shape through which the protein reaches its functional three-dimensional structure. Understanding the mechanisms behind this complex phenomenon and how to direct it, will have an enormous influence in biology and medicine. Recent studies have attributed to mechanical vibrations that extend through the protein a crucial role in controlling structural conformational changes. For example, underdamped low-frequency ($\sim 10^{11}$–10^{12} Hz) collective vibrational modes have been proposed as being responsible for efficiently directing biochemical reactions and biological energy transport. In this contribution we present the results of broad-range Raman spectroscopy measurements made on proteins in different forms, namely crystallized, powdered, and in solution. The use of ultra low frequency (ULF) filters allowed to observe vibrations around 1 THz (30 cm^{-1}), as well as other peaks in the range 0–500 cm^{-1}, that seem to be unassigned in the current literature. Those peaks could be associated to global and/or delocalized vibrational modes. Normal mode numerical calculations were performed on lattice models to investigate on such a possible correlation.

Keywords Terahertz vibrations • Raman spectroscopy • Normal modes • Lattice model • Proteins

3.1 Introduction

The mechanical behavior of proteins is strictly connected to their biological functions. Protein vibrations and their relations with the protein structure (shape) and function are object of investigation since the second half of the last century. Both experimental and computational techniques are currently used to investigate protein dynamics. The former include several spectroscopic techniques, while the latter is basically divided in two approaches, namely molecular dynamics simulation and normal mode analysis (i.e., modal analysis).

Low-frequency vibration modes (near or below 1 THz) play a primary role in driving conformational changes of proteins which are directly related to their biological functions [1–6]. However, low-frequency modes are difficult to detect experimentally, and numerical calculations based on normal mode analysis are thus of help in understanding the role of the low-frequency modes in proteins [7].

This note briefly reports on some experimental and numerical results obtained by the authors on the low-frequency vibrations of proteins, without entering too much in the details. The results are reported here only partially.

3.2 Samples and Methods

Experiments were conducted on protein samples of lysozyme (a well-known enzyme) and Na/K-ATPase (the so-called sodium-potassium pump). As for the former, the samples (Worthington Biochemical Corporation, 2x crystallized, source: hen egg-white) were received and tested in the form of a dialyzed, lyophilized powder. ATPase (Sigma-Aldrich, adenosine 5′-trisphosphatase, source: porcine) was ordered in the form of a lyophilized powder; tests were made in both powdered and rehydrated states. Rehydration was operated by using Milli-Q (Millipore Corporation) purified, deionized water.

Raman spectroscopy measurements [8] were conducted on both proteins. The experimental apparatus consisted in a Raman spectrometer LabRAM HR evolution (Horiba Scientific), mounting a confocal microscope equipped with three

A. Carpinteri • G. Lacidogna (✉) • A. Bassani • G. Piana
Politecnico di Torino, Department of Structural, Geotechnical and Building Engineering, Corso Duca degli Abruzzi 24, 10129, Torino, Italy
e-mail: giuseppe.lacidogna@polito.it

P. Zavattieri et al. (eds.), *Mechanics of Biological Systems, Materials and other topics in Experimental and Applied Mechanics, Volume 4*, Conference Proceedings of the Society for Experimental Mechanics Series, DOI 10.1007/978-3-319-63552-1_3

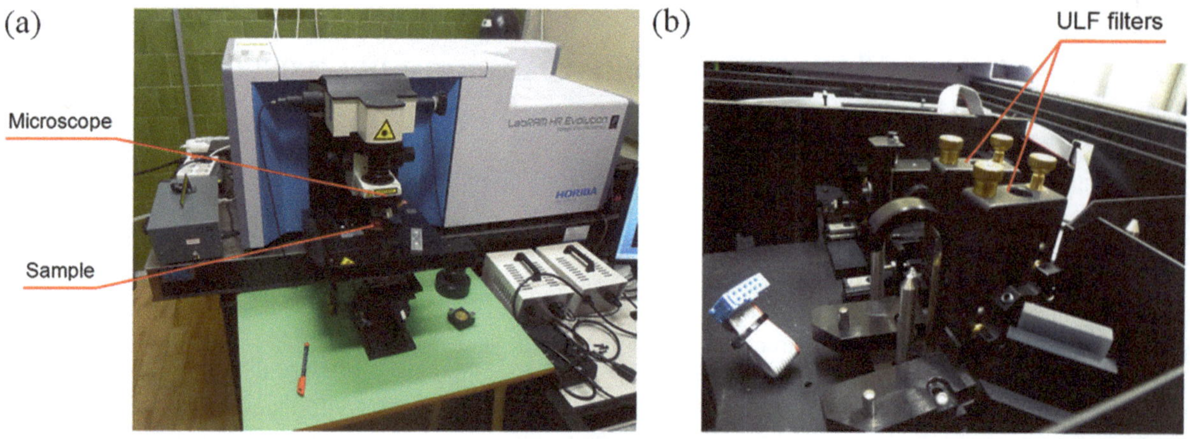

Fig. 3.1 (**a**) Raman spectrometer LabRAM HR evolution equipped with (**b**) a ULF filters module

objectives (20×, 50×, 100×) and coupled to a 633 nm He-Ne laser (Fig. 3.1a). That spectrometer has two diffraction gratings with resolutions equal to 600 lines/mm (<3 cm^{-1}) and to 1800 lines/mm (<1 cm^{-1}). It also mounts ultra low frequency (ULF) filters of the latest generation which allowed investigating the low frequencies (<10–15 THz) (Fig. 3.1b): the ULF module allows getting spectroscopic information in the sub-100 cm^{-1} region, with measurements down to 5 cm^{-1} (0.15 THz) routinely available. The experiments were performed in room conditions, at the temperature of 22 °C, in the High pressure spectroscopy (HPS) laboratory at the Physics Department of Sapienza University in Rome (Italy).

Lastly, normal mode calculations were performed on the crystallographic structure of both proteins. The atom positions of each structure were taken from the RCSB Protein Data Bank [9] (access codes: 4YM8 for lysozyme, 2ZXE for ATPase). Lysozyme and ATPase are composed by 1000 and 10,131 atoms, respectively. All-atom lattice models were built in the commercial finite element code LUSAS [10] and a modal analysis was run. Atoms were simulated by point masses (average atomic mass values were used), while interatomic bonds were replaced by beam elements simulating linear-elastic axial springs (bending and torsion deformations were prevented; average force constant of covalent bonds was used to set the beam axial stiffness).

3.3 Results

It can be theoretically shown, by elementary calculations based on both wave mechanics [11] and structural dynamics [12], that, in terms of orders of magnitude, the vibration frequencies, f, of a body are related to the geometrical dimension, d, involved. Thus, we have: (i) $f \sim 10^{13}$ Hz for localized eigenvibrations of chemical groups/amino acids ($d \sim 10^{-10}$ m); (ii) $f \sim 10^{12}$ Hz for global eigenvibrations of small proteins/delocalized vibrations in extended portions of large proteins ($d \sim 10^{-9}$ m); (iii) $f \sim 10^{11}$ Hz for global eigenvibrations of large proteins ($d \sim 10^{-8}$ m). These results are also confirmed by experiments.

Figure 3.2 shows the broad-range Raman spectra obtained for the lysozyme lyophilized powder sample (Fig. 3.2a) and the ATPase samples, in both the lyophilized (powder) and rehydrated (solution) forms (Fig. 3.2b). In both cases, peaks corresponding to higher frequencies (>500 cm^{-1}) can be recognized and associated to known molecular vibrations. We remark that, in the spectrum of the rehydrated ATPase (Fig. 3.2b, solution), two large peaks corresponding to bending and stretching modes of water are quite marked (specially the stretching mode); the same peaks are much less evident (the stretching, in particular) in the spectrum of the lyophilized ATPase (Fig. 3.2b, powder), while all the other peaks remain in the same position and appear larger. In addition, peaks in the sub-500 cm^{-1} region were found for both proteins. In particular, a large peak at 28 cm^{-1} (0.84 THz) is visible in the lysozyme spectrum (Fig. 3.2a), while several non-negligible peaks <500 cm^{-1} can be seen in the ATPase spectrum (see the lower curve of Fig. 3.2b). To better investigate the lower frequencies, Raman spectra in the 0–500 cm^{-1} region were obtained for both proteins (the results are not shown here). Such low-frequency vibrations were also investigated numerically. Examples are shown hereafter.

Figure 3.3 shows two numerical vibration modes obtained by the finite element simulations. In Fig. 3.3a, a global vibration mode of lysozyme at 0.88 THz (29.3 cm^{-1}) is reported: this corresponds, approximately, to a breathing mode of the whole

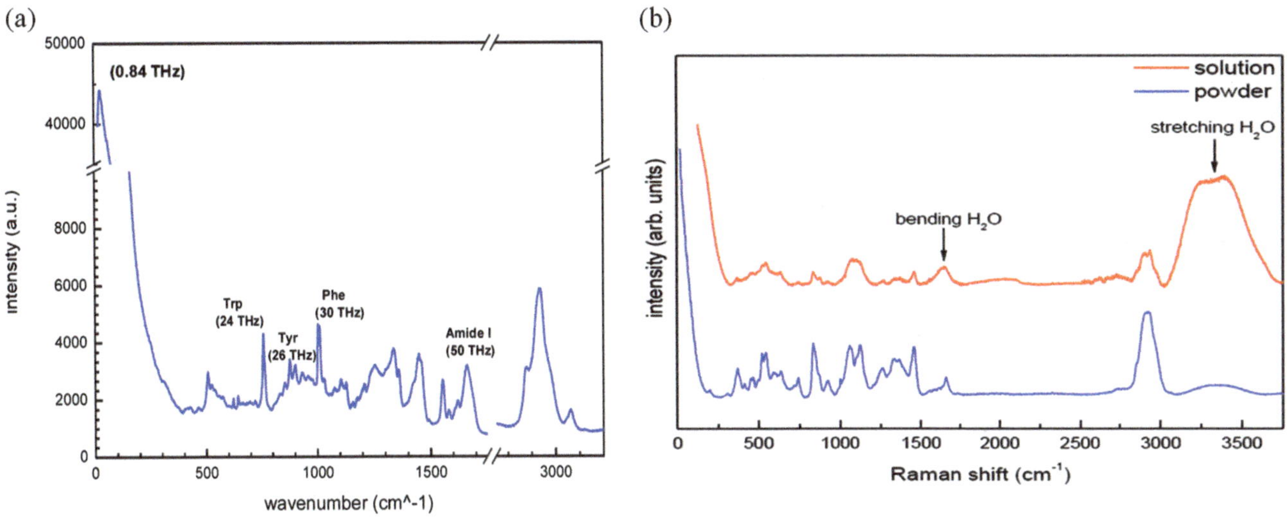

Fig. 3.2 Broad-range Raman spectra of (**a**) lysozyme and (**b**) Na/K-ATPase

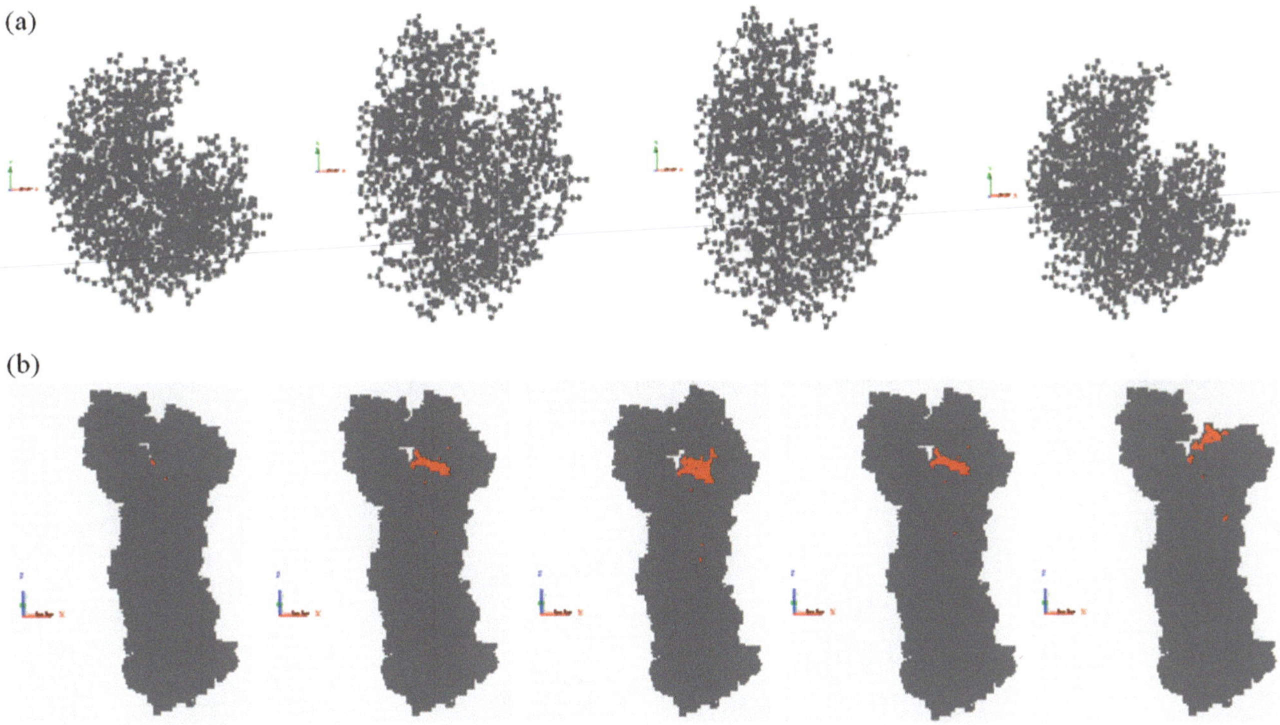

Fig. 3.3 Numerical vibration modes of (**a**) lysozyme at 0.88 THz (29.3 cm^{-1}) and (**b**) Na/K-ATPase at 0.21 THz (7 cm^{-1})

protein. Fig. 3.3b shows a numerical vibration mode of ATPase at 0.21 THz (7 cm^{-1}): this mode involves a delocalized vibration of the top-right horn of the protein.

3.4 Conclusions

Broad-range Raman spectra of lysozyme and Na/K-ATPase were successfully obtained by means of ultra low frequency (ULF) filters mounted on a latest generation spectrometer; this made it possible to investigate also the lower frequencies in the near-THz range. Both proteins were tested in the form of a lyophilized powder; Na/K-ATPase was also tested in solution (rehydrated powder). The Raman spectra obtained for both protein types show peaks in the range $< \sim 500\ \text{cm}^{-1}$ ($\sim$15 THz); in particular: a large peak at 28 cm^{-1} (0.84 THz) was found when analyzing the lysozyme sample; a small peak at 27 cm^{-1} (0.81 THz) and several larger peaks in the range $\sim$ 150–500 cm^{-1} were found for the ATPase samples. These peaks are not assigned in the literature to any structural vibration of chemical groups; in fact, they should correspond to vibrations involving large protein portions (collective or delocalized vibrations) or even the entire protein (global vibrations). Simplified numerical lattice models of the lysozyme and Na/K-ATPase single proteins were therefore implemented in a commercial finite element program. The modal analyses run allowed to find global and/or delocalized vibration modes at frequencies near or below $\sim 10^{12}$ Hz, thus helping in interpreting the experimental results.

Acknowledgments The authors gratefully acknowledge Dr. W. Marchione and Dr. S. Gai of Horiba Scientific for their technical support, as well as Prof. P. Postorino and Dr. C. Fasolato of the Physics Department of Sapienza University (Rome) for performing the Raman measurements.

References

1. Davydov, A.S.: The theory of contraction of proteins under their excitation. J. Theor. Biol. **38**, 559–569 (1973)
2. Deak, J., Chin, H., Lewis, C., Miller, R.: Ultrafast phase grating studies of heme proteins: observation of the low-frequency modes directing functionally important protein motions. J. Phys. Chem. B. **102**, 6621–6634 (1998)
3. Liu, D., et al.: Studies of phononlike low-energy excitations of protein molecules by inelastic X-ray scattering. Phys. Rev. Lett. **101**, 135501 (2008)
4. Acbas, G., Niessen, K.A., Snell, E.H., Markelz, A.G.: Optical measurements of long-range protein vibrations. Nat. Commun. **5**, 3076 (2014)
5. Turton, D.A., Senn, H.M., Harwood, T., Lapthorn, A.J., Ellis, E.M., Wynne, K.: Terahertz underdamped vibrational motion governs protein-ligand binding in solution. Nat. Commun. **5**, 3999 (2014)
6. Niessen, K.A., Xu, M., Markelz, A.G.: Terahertz optical measurements of correlated motions with possible allosteric function. Biophys. Rev. **7**, 201–216 (2015)
7. Nicolaï, A., Delarue, P., Senet, P.: Low-frequency, functional, modes of proteins: all-atom and coarse-grained normal mode analysis. In: Liwo, A. (ed.) Computational Methods to Study the Structure and Dynamics of Biomolecules and Biomolecular . From Bioinformatics to Molecular Quantum Mechanics Processes. Springer, Berlin Heidelberg (2013)
8. Ferraro, J.R., Nakamoto, K., Brown, C.W.: Introductory Raman Spectroscopy, 3rd edn. Elsevier, Burlington (2003)
9. http://www.rcsb.org/pdb
10. http://www.lusas.com/
11. Feynman, R.P., Leighton, R.B., Sands, M.: The Feynman Lectures on Physics. Addison–Wesley, USA (1964)
12. Clough, R.W., Penzien, J.: Dynamics of Structures. McGraw-Hill, New York (1975)

Chapter 4
Digital Image Correlation Study of the Deformation and Functioning of the Human Heart during Open-Heart Surgery

Ayat Soltani, S. Curtze, J. Lahti, K. Järvelä, J. Laurikka, M. Hokka, and V.-T. Kuokkala

Abstract Currently, ultrasound technology is routinely used for monitoring of the left side of the human heart during open-heart surgery. However, this method shows shortcomings in providing accurate information of the right ventricle and atrium. The aim of this paper is to demonstrate how Digital Image Correlation (DIC) can be used to monitor the functioning of the heart during open-heart surgery and potentially overcome some of the shortcomings of ultrasound methods. Being a contact-free method is a major asset from a practical implementation perspective of DIC. In this paper, we present the methodology of the experiment and some preliminary results of a study in which a DIC system was installed in an operating room and image sequences of the heart were taken at three stages of the surgery. We present a procedure for obtaining DIC measurements in this challenging setting, discuss how the data was extracted as well as how the measured values changed during the operation in the context of the surgical stages and interventions performed.

Keywords Open heart surgery • DIC • Deformation • Motion • Biomaterial characterization

4.1 Introduction

Successful open-heart surgery requires accurate monitoring and measurement of myocardial mechanical functions and hemodynamic parameters during the procedure. In particular, mechanical parameters, i.e., the movement, deformation, and overall contractility of the ventricles are major indicators for how well the heart is performing. Abnormal motion and deformation of the ventricles can be an indication of serious problems [1, 2]. In addition to visual evaluation by the clinicians, echocardiography can be used to measure movement, motion, strain, and strain rate of the right ventricle (RV) [1]. However, obtaining quantitative results is not straight-forward due to anatomical and practical reasons, requiring a high skill set from the user whilst an intrusive transesophageal probe is used that in some cases can result in internal bleeding and damage. These problems were discussed in more details by Teske et al. [2]. A special echocardiography based technique that has gained popularity during the past years for analyzing the movement of the heart is the Tissue Doppler Imaging (TDI) [3]. TDI utilizes the Doppler effect to measure the velocity of the myocardium. However, there are limitations associated with this method such as the restriction of measuring only motion parallel to the ultrasound beam, which, once again, puts high user skills requirements on the system operator.

In this work, we are focusing on Digital Image Correlation (DIC) measurements of motion and deformation of the right ventricle (RV), since this is the portion of the heart visible and available for optical photography during the surgery. The practical and anatomical limitations of the ultrasound technology discussed briefly above make Digital Image Correlation a potential candidate for an online, and relatively automated, RV function monitoring method in the future. DIC has already been successfully used in studying many other biological materials, lending further credence to the technique in the present setting. Planca et al. [4] published a comprehensive overview on DIC usage in the biomechanical field. At the most basic level, DIC relies on tracking the movement of points of interest on the target material, from which the deformation can be evaluated. To achieve this, a random speckle pattern with high contrast is typically applied onto the surface of the material. Using one or more cameras, numerous images are acquired of the object during deformation/movement. By sub-dividing the images into smaller subsets, each deformed state image is compared to the reference or "non-deformed" state images one subset at a time. The deformation on the surface of the target can be evaluated using an automated matching algorithm

A. Soltani (✉) • S. Curtze • M. Hokka • V.-T. Kuokkala
Tampere University of Technology, Laboratory of Materials Science, POB 589, FI33101, Tampere, Finland
e-mail: ayat.soltani@tut.fi

J. Lahti • K. Järvelä • J. Laurikka
Tampere University Hospital Heart Center, Tampere University, Tampere, Finland

P. Zavattieri et al. (eds.), *Mechanics of Biological Systems, Materials and other topics in Experimental and Applied Mechanics, Volume 4*, Conference Proceedings of the Society for Experimental Mechanics Series, DOI 10.1007/978-3-319-63552-1_4

between these two, i.e., deformed and reference states. Obtaining high quality DIC results depends on various factors, some of which pose a challenge on our experiments. One such factor is the quality of the contrast pattern on the heart's surface. Commonly, the DIC pattern can be applied using methods such as spray or airbrush [5] or as more recently used, dispersing for instance graphite particles on biological materials [6]. However, none of these methods is applicable to the present study since obviously, toxic materials such as graphite cannot be used on the human heart during the surgical procedure. Even though there are non-toxic, biocompatible inks available that are routinely used in the medical field (e.g., methylene blue), the surface of the heart is wet, which makes it very difficult to use sprays to generate the desired patterns. Another important factor is the perpendicularity of the cameras with respect to the target surface when determining the out of plane movements of the test object. To achieve high quality results, the camera(s) used in DIC experiments should be positioned so that the line of sight forms a 90 degree angle with the target surface. According to the estimations by Hijazi et al. [7], misalignments as small as 2° can lead to strain errors higher than 10^3 $\mu\varepsilon$. Similarly, Hiddadi et al. [8] investigated the effects of out of plane motion on the accuracy of the DIC results. He concluded that the errors are linearly proportional to the amount of out of plane displacement. Both considerations are of relevance to this paper because the distance from the cameras to the target surface is not exactly known during the imaging and calibration procedures. Furthermore, the patient's respiration and the movement of the heart during each beat will results in an out of plane movement of the region of interest. Finally, the cameras are placed at angle that is smaller than 90 degrees to operating table. In spite of these potential error sources, DIC provides additional information of the heart's deformation and functions compared to the conventionally used monitoring methods in the medical field. In our previous work, [9–11] we have shown that DIC can be used to measure motion and deformation of the heart during surgery. In this paper, we will show how the biomechanical response of the heart changes during a surgical procedure and present a preliminary evaluation of the sensitivity of selected values and parameters in describing changes in the overall response of the heart.

4.2 Experimental

The images for this study were acquired during an open-heart surgery using the DIC setup after installation in the operating theater at the Tampere University Hospital Heart Center in Tampere, Finland. A picture and a schematic of this setup are shown in Fig. 4.1a, b, respectively. The DIC setup comprises two 5-megapixel E-lite cameras equipped with 50 mm Nikon lenses. The equipment was installed above the surgery table, overlooking the patient's chest from an angled view. Typically, a perpendicular view is the optimum geometrical alignment of the DIC cameras towards the target surface. However, the sterile field above the patient is under strict regulations, making it impossible to position the cameras directly on top of the surgery table. Also readjusting the cameras during the surgical procedure is restricted, since any interference with the sterile

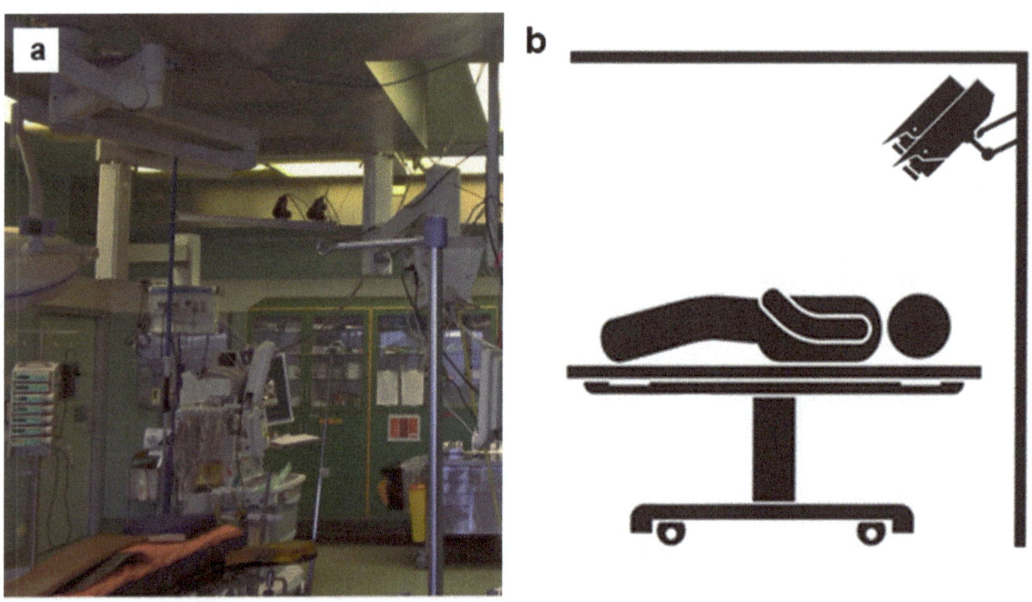

Fig. 4.1 (**a**) A picture of the DIC setup installed in the operating theater and (**b**) a schematic picture of the DIC setup

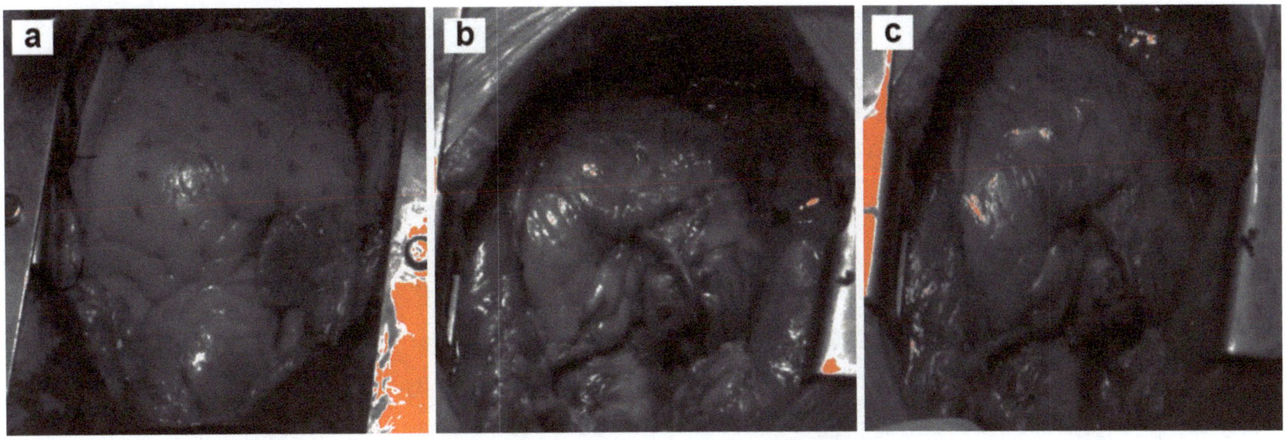

Fig. 4.2 Contrast patterns for **(a)** the native stage, **(b)** the weaning stage, and **(c)** the drug stage

Table 4.1 Parameters and calibration errors for the DIC calculations

Subset size	79 pixels
Step size	10 pixels
Pixel size	~0.003 mm
Calibration error	~0.1 pixel

area around the patient is prohibited. For these reasons, calibration was done after the surgery by holding the calibration plate approximately at the position where the heart was placed during the surgery. The calibration was then transferred to the surface of the heart numerically using an iterative procedure described in ref. [12].

The natural texture of the heart does not provide an ideal contrast for high quality DIC studies [9, 10, 13]. This is the case for most of the heart surgery patients who typically have a thin layer of fat covering the myocardium giving it a uniform white/yellowish color. To somewhat mitigate this problem, the surgeons applied a few random dots on the heart's myocardium (the most outer surface) using a sterile, non-toxic surgical marker that is routinely used in the medical field. The pattern was applied to the heart just before recording of the imaging sequences. Applying the pattern during the latter stages of the surgery appeared to be somewhat more challenging, most likely due to the surface of the myocardium drying out throughout the procedure, and therefore the quality of the pattern was worse as shown in Fig. 4.2. During the surgery, the *native stage* images were obtained after opening the pericardium, the *weaning stage* images after performing the surgical repairs was finished and the spontaneous function of the heart had been stabilized (typically several hours after the native stage), and the *drug stage* images after a theophylline bolus, only a couple of minutes after the weaning stage images. Theophylline improves the activity of the RV and the hypothesis was that the changes are visible in the DIC analysis.

During each imaging stage of the surgery, approximately 200 images were obtained at a sampling rate of about 18 Hz, i.e., around 600 images for each surgery in total. In this paper, we present the processed data covering these three stages. The recorded images were processed and analyzed using the Davis software v8.3 by LaVision. The calculation parameters are shown in Table 4.1.

Several images of a 3D calibration target were acquired after the surgical procedure to obtain a stereo calibration for the DIC setup. The calibration error presented in Table 4.1 was calculated as the root mean square difference between the positions of the points in the calibration images and their known positions in the real world. The images were cropped down to 2000 by 1700 pixel from the original 5 MPixels to increase the image acquisition frequency from about 10 to approximately 18 Hz. The displacement vectors were calculated using the sum of differential routine in the Davis software, where vector displacement is always determined from two consecutive images. This calculation routine gradually accumulates errors, which can become substantial towards the end of a long image sequence. Processing the images for each cardiac cycle individually and using the software to stitch them back together significantly alleviated this problem.

4.3 Results and Discussion

Engineering strain was calculated using a virtual extensometer across the surface of the heart, while full field data was obtained from a user specified region of interest (ROI), manually defining the visible edges of the right ventricle and thereby excluding the outside regions from the analysis. Both the virtual extensometer and the ROI used in this study are shown in

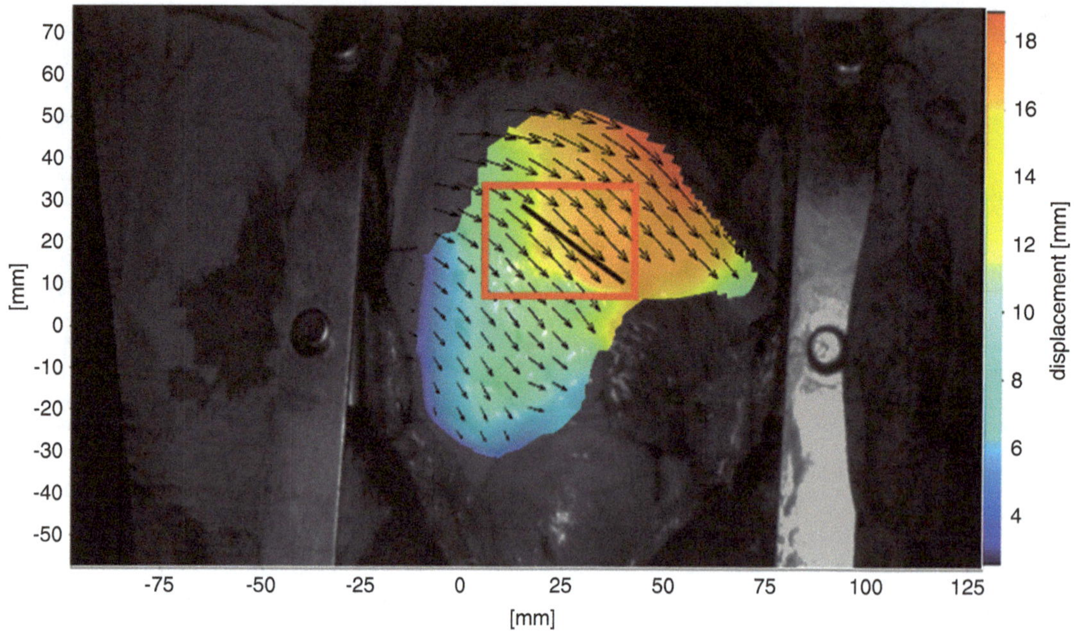

Fig. 4.3 An image of the heart overlaid with the displacement vectors and a displacement heat map. The thick *black line* shows the length and direction of the virtual extensometer, and the *red rectangle* represents the region of the interest from which strain data was extracted

Fig. 4.3. The virtual extensometer was aligned roughly along the direction of the maximum elongation, i.e., in the principal direction of the displacement vectors, which should also coincide with the alignment of the muscle fibers.

4.3.1 Principal Strains

Figure 4.4 shows the maximum and minimum principal strains calculated for the three stages. In Fig. 4.4a, the highest maximum principal strain levels for the native stage average roughly around 40% for most heart beats, reaching just above 50% for two beats. The peak value of the maximum principle strain drops to just around 30% after the surgical procedure in the weaning stage, showing a reduction in the tensional strain on the surfaces of the heart. Furthermore, the shape of each curve representing a single cardiac cycle becomes less distinct in this stage as compared to the native stage. In the native stage, the curve shape reveals some characteristic features, such as a clear step on both the rising and the tailing edges of each peak. These are not clearly observable in the data obtained from the weaning stage. In the present dataset, the theophylline bolus shows only a weak effect on the maximum amplitude level for the maximum principle strain, but the shape of the pulses start to become more distinct and clearly resemble the pulse shapes observed during the native stage. The correlation of the minimum principal stains with the maximum principle strains and at the same time with the three stages is very interesting to notice. Whilst the maximum principal stain amplitude decreased from the native stage to the weaning stage, the minimum principal strain amplitude increases (to negative, i.e., compressive values in excess of −25%), whilst the stepwise character of the curve becomes more manifested in the minimum principal stain, indicating a clear change in the deformation characteristics. The increased compressive strains of the weaning stage decrease back to approximately the same amplitude level as observed for the native stage after the theophylline bolus, but the "shoulder levels" remain elevated.

4.3.2 Vector Components

Figure 4.5 shows the displacements in X, Y, and Z directions and the overall length of the displacement vectors for each stage. Using a right-handed Cartesian coordinate system with the X-direction aligned with the horizontal edge, positive direction to the right, and positive Y upward along the vertical edge, the Vy values were originally mostly negative, but in Fig. 4.5 the Y direction was inverted to make the comparisons easier.

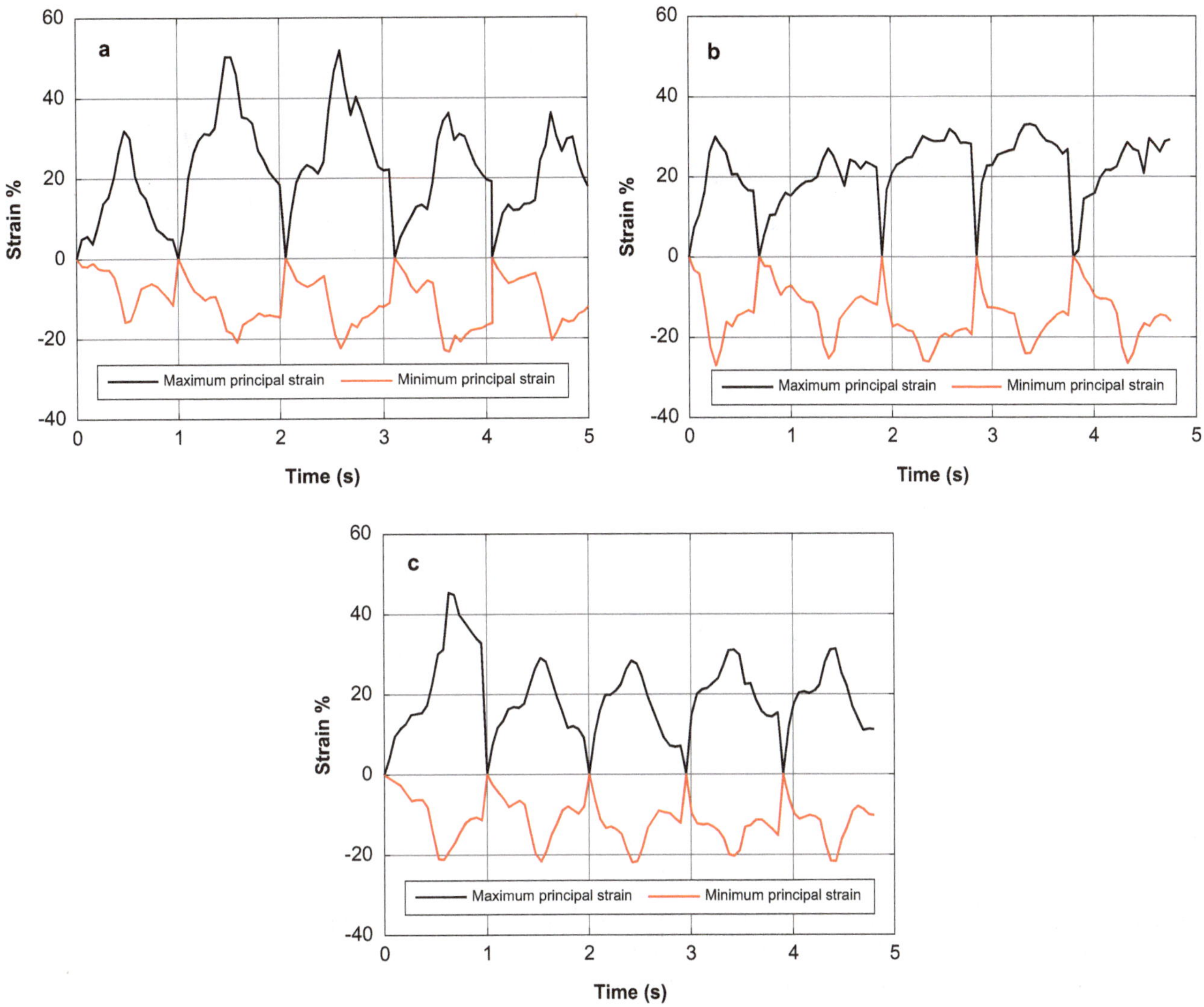

Fig. 4.4 The maximum and minimum principal strains for (**a**) the native stage, (**b**) the weaning stage, and (**c**) the drug stage

At all stages, the pulses for vector length V_{lenght}, V_x and V_y retain the same overall shape. Each pulse shows an increase up to the half/two thirds of the peak value, then drops slightly before picking up again to reach the highest value. For all the vector components, the pulses get somewhat wider in the weaning stage. Especially the intermediate maxima or plateaus at the rising edge of the pulses are much longer than in the native stage. The injection of the drug results in each component becoming somewhat sharper, resembling the native stage. However, the intermediate maxima are still longer than in the native stage, but shorter than in the weaning stage. Comparing the native stage to the weaning stage, the peak deformation amplitude of each heart beat drops by several millimeters. This drop can be observed for all vector components. Overall, the injection of the theophylline seems to slightly increase the peak deformation value level, although this effect is only seen in some pulses/heart beat cycles. In general, there's a modulation superimposed on the heart's deformation characteristics that can be attributed to the ventilation of the patient and the heart-lung interaction, making comparison of single heart beats somewhat difficult.

V_z behaves very differently from the other vector components. Even in the native stage, the pulses for each cycle vary vastly from one another, both in shape and value. In the native stage, the V_z values fluctuate between 10 mm and as low as −5 mm in the third cycle. Here it should be noted that the zero-level is determined by the calibration routine and therefore somewhat arbitrary. The major level change can be attributed to the ventilation cycle due to which the patient's chest rises and lowers, which covers about 6–7 heart beats in duration (about 6–7 respiratory/ventilation cycles per minute). In the weaning stage, the V_z graph changes so much that it is very difficult to distinguish each individual pulse. The peak values for V_z decrease by around 5 mm compared to the native stage, shifting between 5 and −5 mm. Finally, similar to the other

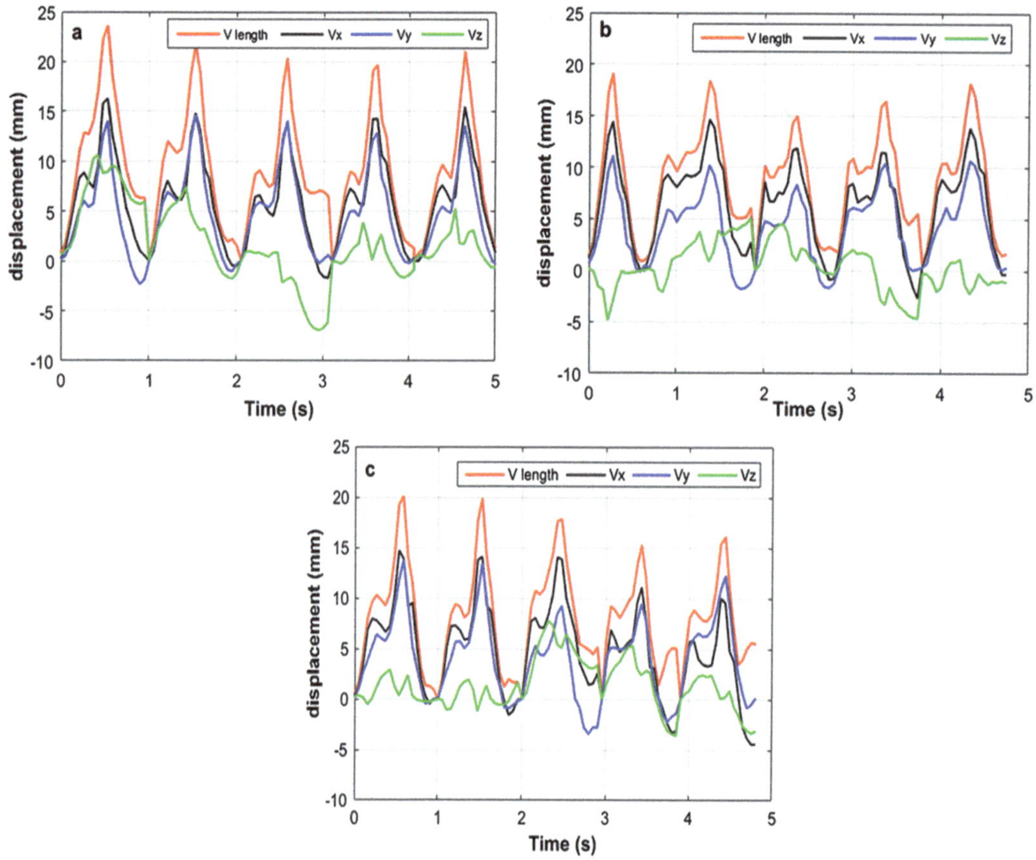

Fig. 4.5 The length of the displacement vector and its components for (**a**) the native stage, (**b**) the weaning stage, (**c**) the drug stage

vector components, introducing the drug leads to an increase in peak values for V_z and the pulses become somewhat more regular shaped, resembling the behavior observed in the native stage. During all stages, the major deformation occurs in the X-Y plane, while the major fraction of Z movement can apparently be attributed to ventilation.

4.3.3 Fractional Shortening

Fractional shortening (FS) is typically used to assess the ability of the heart to pump blood into circulation [14]. Conventionally, FS is measured using ultrasound techniques. Its physical interpretation is the relative change in circumferential dimensions of the ventricle's inner dimensions during one heartbeat, and the exact determination depends on the ultrasound projection. Using this concept as the basis, a similar value in the longitudinal direction can be calculated from the data acquired by DIC. For this purpose, the length of the line shown in Fig. 4.3 is measured for each image/frame and normalized by the maximum length of the line for each cycle. The maximum length corresponds to the point in time in each heartbeat where the heart is at its maximum dilation (end-diastole). FS was calculated from the DIC data using Eq. 4.1:

$$FS(t) = \left(1 - \frac{L(t)}{L_{max}}\right) \times 100\% = \frac{L_{max} - L(t)}{L_{max}} \times 100\% \qquad (4.1)$$

The calculated FS values for the three stages of the surgical procedure are shown in Fig. 4.6. In the native stage, the peak FS reaches values as high as 26% for one heart-beat and around 13% at the lowest. In the weaning stage the peak values level around 8–10% showing a clear drop in the overall FS values, and in the drug stage the peak values slightly increase to values between 10–13%. The pulse shape for the native stage is different from the two other stages. In the native stage, the pulses show a step on the rising and tailing slopes of each pulse, whereas these steps are not so evident in the later stages.

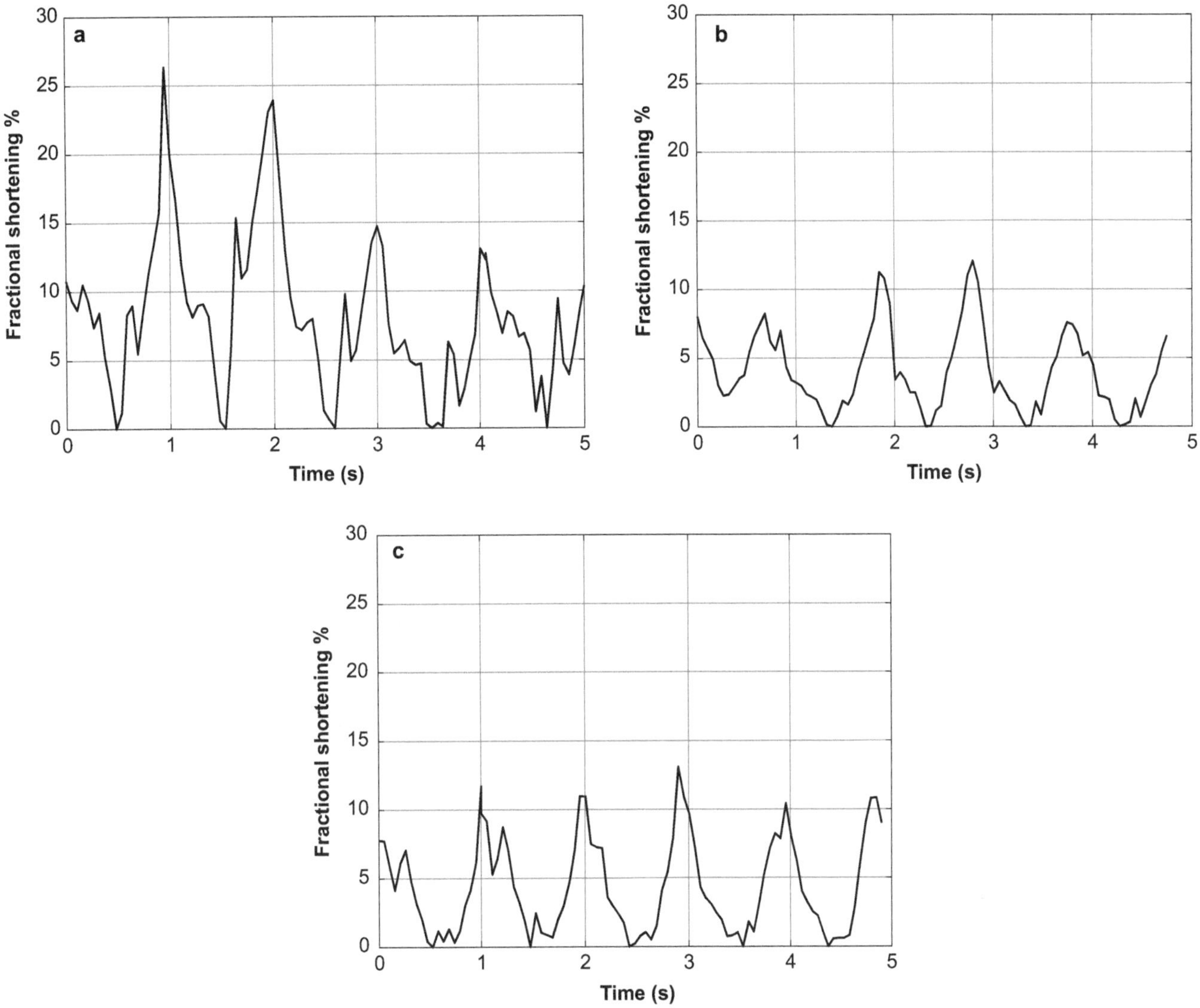

Fig. 4.6 Fractional shortening calculated from the DIC data for (**a**) the native stage, (**b**) the weaning stage, and (**c**) the drug stage

4.3.4 Strain Rate

Recently, strain and strain rate have proven to be useful concepts for assessing and monitoring myocardial functionalities [15, 16, 17]. Since the FS is mathematically identical with compressive engineering strain, it can be used to calculate the strain rate. Compared to FS, the strain rate values shown in Fig. 4.7 show much more fluctuations and the heart-beats do not have a clear pulse-like shape. In the native stage, the strain rates are rather high and the peak values reach 200 s^{-1}. By comparison, in the weaning state the maximum strain rates drop to about 50 s^{-1} and the average minimum values decrease in amplitude to about -50 s^{-1}. The strain rates in the drug stage are comparable to those observed in the weaning stage, with a small decrease in the level of tensile (dilative) strain rates.

Some clear changes in the biomechanical response and mechanical behavior of the heart were observed after the surgical procedure and after the drug injection. The principle strains and the displacement vector components show relatively strong changes with the injection of the drug and the shape of the pulse and the amplitudes change towards that observed at the native stage. In the other measured strain parameters, the changes were not that obvious in this limited experimental study of one patient and only for 5 heart beats. In general, the amplitudes of the different deformation pulses depend on the respiration, which cyclically moves the patient's chest up and down. The effect of respiration on the deformation and motion of the heart is not obvious in the limited data used in this work, but preliminary analysis of a much longer image sequences have shown modulation in the amplitude values that resembles that observed in the ultrasound measurements.

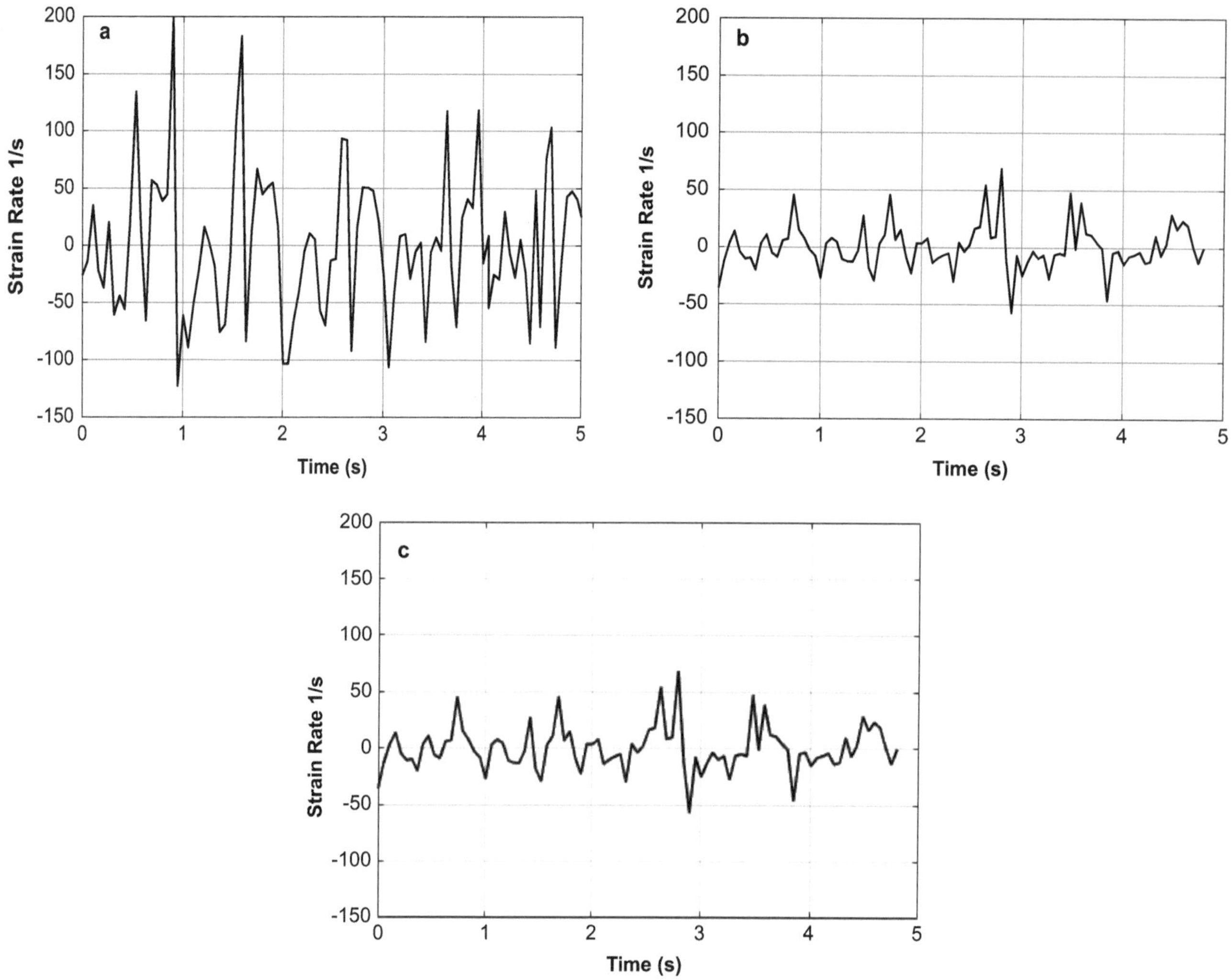

Fig. 4.7 Engineering strain rate as a function of time for (**a**) native stage, (**b**) wean stage, and (**c**) drug stage

4.4 Conclusions

A 3D DIC setup was installed in an operating theater, and image sequences of the patient's heart were acquired during three stages of an open heart surgery. These images were analyzed using the Davis software by LaVision. Various measures describing the biomechanical response of the heart were extracted to study the difference in the heart's behavior during the surgery. The surgical procedure caused a noticeable change in the biomechanical response and activity of the heart. Both the peak deformation values and the pulse shapes in all of the data types presented in this paper showed clearly identifiable changes. These changes include the drop in the peak values and the pulses becoming wider right after the surgical repairs were finished and the patient was taken from cardiopulmonary bypass support. The interrelated effects of the theophylline bolus, however, were not completely obvious in all strain and deformation values studied in this work. For instance, the strain measurements in the drug stage indicated an increase in the peak values while strain rate showed no noticeable change, or rather a slight decrease during the filling. However, the strain rate requires time differentiation of the strain signal, which produced rather noisy signals due to the rather low image acquisition rate compared to the amount of deformation. The initial work presented in this paper only covers one patient and five heart beats for each stage. Therefore, it is clear that more studies are required to confirm the conclusions of this work and better understand the correlations. The future work also includes improving the imaging conditions to obtain better contrast and the use of higher image acquisition rates. The DIC measurements should be compared to simultaneous TDI measurements for better understanding of the overall mechanical behavior and functionality of the heart.

References

1. La Gerche, A., Robberecht, C., Kuiperi, C., Nuyens, D., Willems, R., de Ravel, T., Matthijs, G., Heidbüchel, H.: Lower than expected desmosomal gene mutation prevalence in endurance athletes with complex ventricular arrhythmias of right ventricular origin. Heart. **96**(16), 1268–1274 (2010)
2. Teske, A., Robberecht, C., Kuiperi, C., Nuyens, D., Willems, R., de Ravel, T., Matthijs, G., Heidbüchel, H.: Echocardiographic tissue deformation imaging quantifies abnormal regional right ventricular function in arrhythmogenic right ventricular dysplasia/cardiomyopathy. J. Am. Soc. Echocardiogr. **22**, 920–927 (2009)
3. Sutherland, G., Kukulski, T., Uwe Voight, J., Dhooge, J.: Tissue Doppler echocardiography. Echocardiography. **16**, 509–520 (1999)
4. Planca, M., Tozzi, G., Cristofolini, L.: The use of digital image correlation in the biomechanical area: a review. Int. Biomech. **3**(1), 1–21 (2015)
5. Lionello, G., Cristofolini, L.: A practical approach to optimizing the preparation of speckle patterns for digital-image correlation. Meas. Sci. Technol. **25**(10), 107001 (2014)
6. Barranger, Y., Doumalin, P., Dupré, J., Germaneau, A.: Digital image correlation accuracy: influence of kind of speckle and recording setup. In Germaneau (2010)
7. Hijazi, A., Friedl, A., Kahler, C.: Influence of Camera's optical Axis non-perpendicularity on measurement accuracy of two-dimensional digital image correlation. Jordan J. Mech. Indust. Eng. **5**(4), 373–382 (2011)
8. Hiddadi, H., Belhabib, S.: Use of rigid-body motion for the investigation and estimation of the measurement errors related to digital image correlation technique. Opt. Lasers Eng. **46**(2), 185–196 (2008)
9. Hokka, M., Mirow, N., Vogt, N. and Kuokkala, V.-T.: DIC Measurements of the Human Heart during Cardiopulmonary Bypass Surgery. In: SEM annual conference, Costa Mesa (2015)
10. Hokka, M., Mirow, N., Nagel, H., Irqsusi, M., Vogt, S., Kuokkala, V.: In-vivo deformation measurements of the human heart by 3D digital image correlation. J. Biomech. **48**, 2217–2220 (2015)
11. Mirow, N., Hokka, M., Nagel, H., Vogt, S., Irqsusi, M., Kuokkala, V.-T., Moosdorf, R.: Intraoperative stereoscopic strain analysis of right ventricular wall-motion changes after coronary bypass surgery. Biomech. Eng. **3**, (2015)
12. Hokka, M., Mirow, N., Nagel, H., Vogt, S., Kuokkala, V.-T.: DIC measurements of the human heart during cardiopulmonary bypass surgery. Mech. Biol. Syst. Mat. **6**, 51–59 (2016)
13. Dandel, M., Lehmkuhl, H., Knosalla, C., Suramelashvili, N., Hetzer, R.: Strain and strain rate imaging by echocardiography–basic concepts and clinical applicability. Curr. Cardiol. Rev. **5**, 133–148 (2009)
14. Lewis, R., Sandler, H.: Relationship between changes in left ventricular dimensions and the ejection fraction in man. Circulation. **44**(4), 548–557 (1971)
15. Denault, A.Y., Couture, P., Beaulieu, Y., Haddad, F., Deschamps, A., Nozza, A., Pagé, P., Tardif, J.-C., Lambert, J.: Ventricular depression after cardiopulmonary bypass for Valvular surgery. J. Cardiothorac. Vasc. Anesth. **29**(4), 836–844 (2015)
16. Voelkel, N.F., Quaife, R.A., Leinwand, L.A., Barst, R.J., McGoon, M.D., Meldrum, D.R., Dupuis, J., Long, C.S., Rubin, L.J., Smart, F.W., Suzuki, Y.J., Gladwin, M., Denholm, E.M., Gail, D.B.: Right ventricular function and failure. Circulation. **114**, 1883–1891 (2006)
17. Marwick, T.: Measurement of strain and strain rate by echocardiography: ready for prime time? J. Am. Coll. Cardiol. **7**, 1313–1327 (2006)

Chapter 5
Anisotropic Conductivity Tensor for Irreversible Electroporation Simulations

Nicholas LaBarbera and Corina Drapaca

Abstract In the past few years, interest has drastically increased in using surgically inserted electrodes to ablate cancer cells. The treatment is referred to as irreversible electroporation (IRE) and has the advantage of being a minimally invasive procedure that can be used to treat tumors while minimizing damage to surrounding tissue and preserving blood vessels. However, treatment planning is required to ensure the electrodes are placed in the correct location and at the proper voltages to ensure all cancer cells are killed while causing the least amount of damage to healthy cells. Computer simulations are used to develop individual treatment plans. Currently, computer simulations for treatment planning have used an isotropic conductivity that increased as the tissue was electroporated. In this work, we propose using a conductivity tensor in which the conductivity increases more in the direction of the electrical field, and results in an anisotropic conductivity tensor. This slightly alters the predicted ablation zone, and could lead to more accurate simulations.

Keywords Irreversible electroporation • Numerical modeling • Tumor ablation • Treatment planning

5.1 Introduction

Electroporation is an electrically-driven biological process that uses externally applied electric fields to briefly open the nano-sized pores in a cell's membrane. On one hand, electroporation can be used to transport drugs and genetic materials into cells [3]. On the other hand, a strong enough electrical field can irreversibly damage the cell's membrane and cause cell death. This process is called irreversible electroproation (IRE), and has gained interest as a possible tumor ablation method [7].

A cell's viability is directly related to the electrical field is exposed to. Each cell is characterized by a critical value so that when the cell is exposed to an electrical field above this critical value it loses its ability to reseal the pores in its membrane and eventually dies. Any electrical field below this critical value, either does not form pores or the pores close when the electrical field is removed. The critical value depends of various factors such as tissue type, tumor stage, electrode configuration and size, and the direction of the applied electric field. The success of an IRE-based therapy for tumors relies on this critical value and thus knowledge about the optimal configuration and size of the electrodes as well as the applied voltage needed to completely ablate the tumor while minimizing the damage to healthy tissue is relevant to clinicians.

Mathematical models of electroporation and corresponding numerical simulations able to predict the electric field distribution in a tissue of interest for various geometries and positions of electrodes could play an important, complementary role in the design of individual treatment protocols [10]. In particular, it is well established in the literature that electroporation can affect the conductivity of tissue [5, 12]. The large increase in conductivity is a result of the membrane's pores opening during electroporation [6]. Corovic et al. [2] recommended to incorporate in mathematical models a conductivity that is dependent on the electrical field strength to increase accuracy. Experiments performed by Mezeme et al. suggest that electroporation does not always isotropically increase the conductivity of the tissue, but conductivity is increased more in the direction tangent to the electric fieldz [9]. Their experiments determined the conductivity of electroporated chicken livers by using magentic resonance electrical impedance tomography. The result of an anistropic conductivity tensor can be explained by the pores forming in the cell membrane with a bias towards the direction of the electrical field [13]. This is because the voltage drop across the cell membrane is not equal around the entire cell, but is largest where the cell membrane is perpendicular to electrical field [13].

N. LaBarbera (✉) • C. Drapaca
Engineering Science and Mechanics, The Pennsylvania State University, State College, PA, USA;
e-mail: nal5047@psu.edu

P. Zavattieri et al. (eds.), *Mechanics of Biological Systems, Materials and other topics in Experimental and Applied Mechanics, Volume 4*, Conference Proceedings of the Society for Experimental Mechanics Series, DOI 10.1007/978-3-319-63552-1_5

Therefore, the aim of this paper is to propose a mathematical model that incorporates a tissue's conductivity increasing more in the direction of the electrical field. By performing numerical simulations and comparing isotropic and anisotropic varying conductivities, we evaluate the importance of including anisotropically varying conductivity into a simulation of treatment planning.

5.2 Mathematical Model

In numerical simulations for treatment planning, it is assumed that the applied direct current electric field is at steady state [4]. The electrical potential ϕ satisfies the differential form of Gauss's law at steady state:

$$\nabla \cdot (\sigma \nabla \phi) = 0 \tag{5.1}$$

where σ is tissue's conductivity. Then the electric field E is given by:$E = -\nabla \phi$. If σ is a constant, then equation (5.1) becomes the well-known Laplace equation. However, the conductivity of tissue is not in general constant but increases during electroporation [5], and therefore, in this paper, we model the conductivity as a function of the electric field, $\sigma = \sigma(E)$. We model the isotropic varying conductivity as a sigmoid Gompertz curve[1] [2, 8, 11]:

$$\sigma(\|E\|) = \sigma_0 + (\sigma_{max} - \sigma_0) \cdot exp[-A \cdot exp(-B \cdot \|E\|)] \tag{5.2}$$

where A and B are unitless coefficients, σ_0 is the base conductivity before electroporation, σ_{max} is the maximum conductivity a tissue can achieve after electroporation, and $\|E\|$ is the l^2- norm of the electrical field. By definition, the l^2- norm of the electrical field is $\|E\| = \sqrt{E_x^2 + E_y^2 + E_z^2}$. The values of A and B are found by fitting the curve for conductivity to experimental data. In Eq. 5.2, σ is only a function of the electrical field's magnitude (Fig. 5.1).

The anisotropic varying conductivity σ is a tensor that depends on the electrical field's direction and magnitude. The 2D Cartesian conductivity tensor used in this paper is given by

$$\sigma = \begin{pmatrix} \sigma_t \cos^2(\theta) + \sigma_n \sin^2(\theta) & (\sigma_n - \sigma_t)\cos(\theta)\sin(\theta) \\ (\sigma_n - \sigma_t)\cos(\theta)\sin(\theta) & \sigma_n \cos^2(\theta) + \sigma_t \sin^2(\theta) \end{pmatrix} \tag{5.3}$$

where σ_t is a function representing the conductivity in the direction of the electrical field, σ_n a function modeling the conductivity in the direction normal to the electrical field, and θ is the angle between the $x-$ axis and the direction of the electrical field.

We solve Eq. (5.1) with σ given by (5.2) and, respectively, (5.3). The boundary conditions used in our numerical simulations are mentioned in the next section.

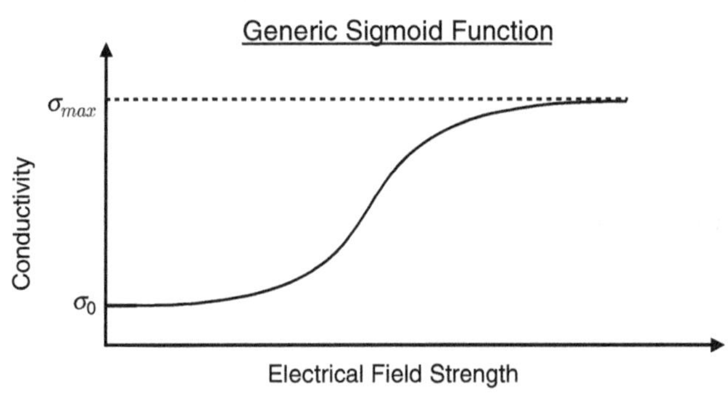

Fig. 5.1 Plot of a generic sigmoid function. In the plot σ_0 represents the conductivity before electroporation and σ_{max} is the conductivity after electroporation

[1]Interestingly, the Gompertz curve was used successfully to fit experimental data of tumors' growth.

5.3 Numerical Model

All simulations were performed using the commercial finite element software COMSOL. COMSOL was chosen as the simulation tool because numerous other groups used COMSOL for IRE simulations [2, 14]. To handle non-linearities, Newton's method was utilized and a direct solver was used for matrix inversion.

5.3.1 *Model Geometry*

We used two different geometries. The first is a 2D monopolar geometry and the second is a 3D bipolar one.

The 2D monopolar case is based off the experiments and simulations performed by Castellvi et al. [1]. A schematic of the geometry is shown in Fig. 5.2. The boundary conditions are as follows. The left electrode is set to a voltage V_0, so:

$$\phi = V_0 \tag{5.4}$$

on the left electrode surface. The right electrode is set to ground, and thus:

$$\phi = 0, \tag{5.5}$$

on the right electrode. Lastly, the outer surface and electrode sleeves are electrically insulated:

$$\frac{\partial \phi}{\partial n} = 0 \tag{5.6}$$

For the 3D bipolar case, the domain was a sphere of 60 mm diameter. The electrode is cylindrical and has a diameter of 1.5 mm and a length of 40.5 mm. The last 7 mm of the electrode are set to the positive voltage. The next 7 mm of the electrode is insulated, followed by 7 mm of the electrode which is set to ground. Lastly, the remaining portion of the electrode and the outer sphere are insulated electrically. A schematic of this case is shown in Fig. 5.3.

5.3.2 *Model Parameters*

For the isotropic conductivity we used:

$$\sigma(\|E\|) = 0.03 + 0.35 * \exp(-\exp(-0.01(\|E\| - 250)), \tag{5.7}$$

which was found by Castellvi et al. [1] by fitting the curve to experimental data.

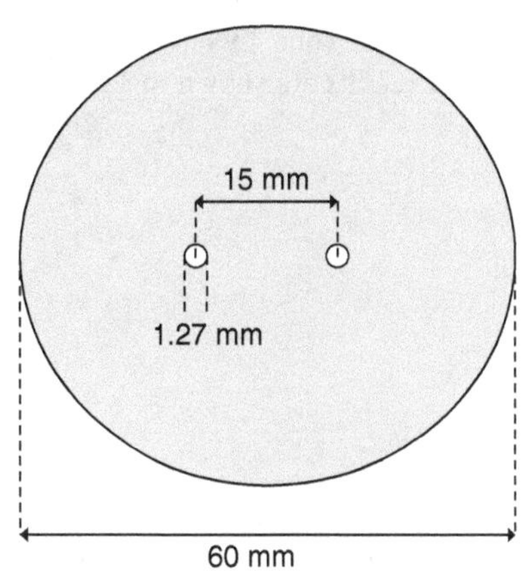

Fig. 5.2 Schematic of the simulation perform for a single tissue. Two needle electrodes with diameter 1 mm are placed 8 mm apart. The entire simulation domain is a 60 mm diameter circle

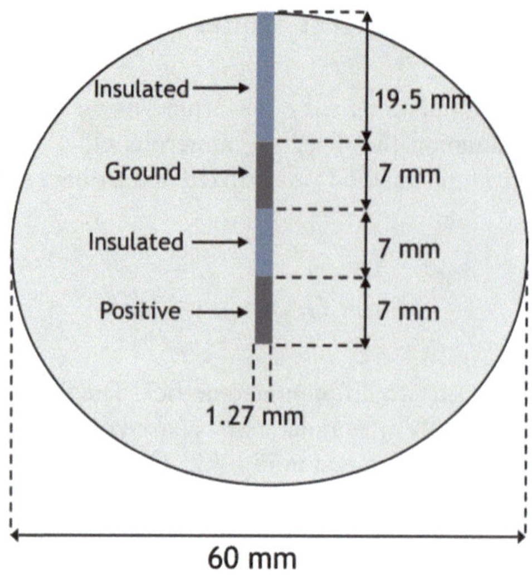

Fig. 5.3 Plot of the schematic for the bipolar test case

For the anisotropic conductivity, the conductivity in the direction of the electrical field was taken to be:

$$\sigma_t(\|E\|) = \sigma(\|E\|) = 0.03 + 0.35 * \exp(-\exp(-0.01(\|E\| - 250)) \tag{5.8}$$

The conductivity in the direction perpendicular to the direction of the electrical field was:

$$\sigma_n = (1 - \sigma_\Delta)\sigma_t, \tag{5.9}$$

where

$$\sigma_\Delta(\|E\|) = 0.30 * \exp(-\exp(-0.01 * (\|E\| - 250))) \tag{5.10}$$

was found using the experiments reported in [9].

A volume was considered irreversibly electroporated if the magnitude of the electrical field was 184 V/cm or greater. This value was determined experimentally in [1].

5.3.3 *Results*

For both the monopolar and the bipolar simulations, the anisotropic simulations predicted a smaller ablation. The difference was larger for the bipolar simulations. This is in accordance with the results reported in [14], where it was found that the isotropic formulation for bipolar probe simulations over-predicts ablation zones more than the isotropic monopolar probe simulations. Thus, by having a larger change in volume for the bipolar case, the results are following the experimental trend.

Our results are shown in Table 5.1 for the monopolar case, and in Table 5.2 for the bipolar case.

Table 5.1 Results for the 2D monopolar simulation

	Isotropic IRE area(mm^2)	Anistropic IRE area(mm^2)	Percent difference
$V_0 = 500$ V	225	221	1.8
$V_0 = 750$ V	396	384	3.1
$V_0 = 1000$ V	562	538	4.5
$V_0 = 1250$ V	724	689	5.1

Table 5.2 Results for the 3D bipolar simulation

	Isotropic IRE vol.(cm^3)	Anistropic IRE vol.(cm^3)	Percent difference
$V_0 = 500$ V	2.49	2.33	6.4
$V_0 = 750$ V	5.27	4.67	11.8
$V_0 = 1000$ V	7.89	7.06	12.8
$V_0 = 1250$ V	10.60	9.38	13.0

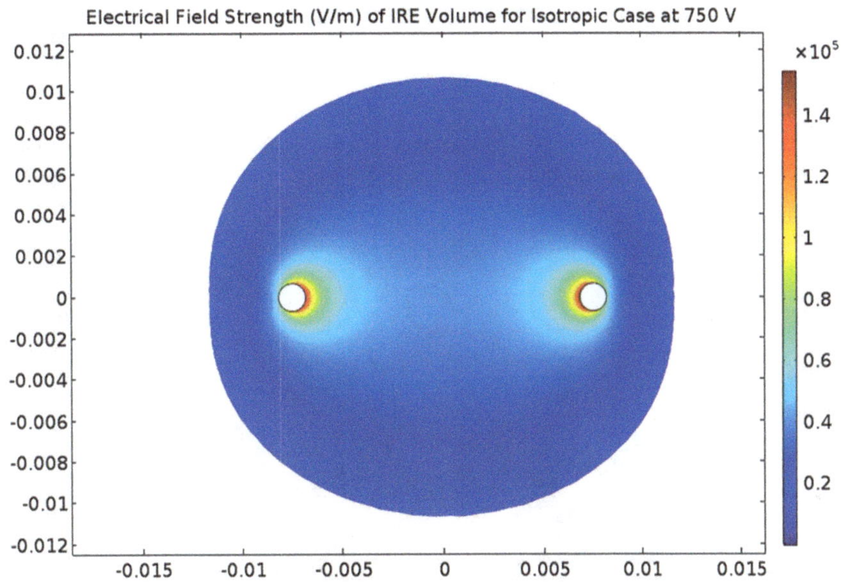

Fig. 5.4 Plot showing the IRE volume for the isotropic simulation of 750 V. Colors represent electrical field strength. *White regions* correspond to areas in the domain where the electrical field was not strong enough to induce cell death from IRE

5.4 Conclusion

In this paper we presented numerical simulations of electroporation for isotropic and anisotropic varying conductivity of the tissue in two cases: 2D monopolar and 3D bipolar geometries. Our preliminary results for the anisotropic formulation for the conductivity tensor seem promising. They show a slight decreasing in the predicted volume of ablated domain, and could be a possible explanation for the over-prediction seen in isotropic formulations (Figs. 5.5 and 5.6).

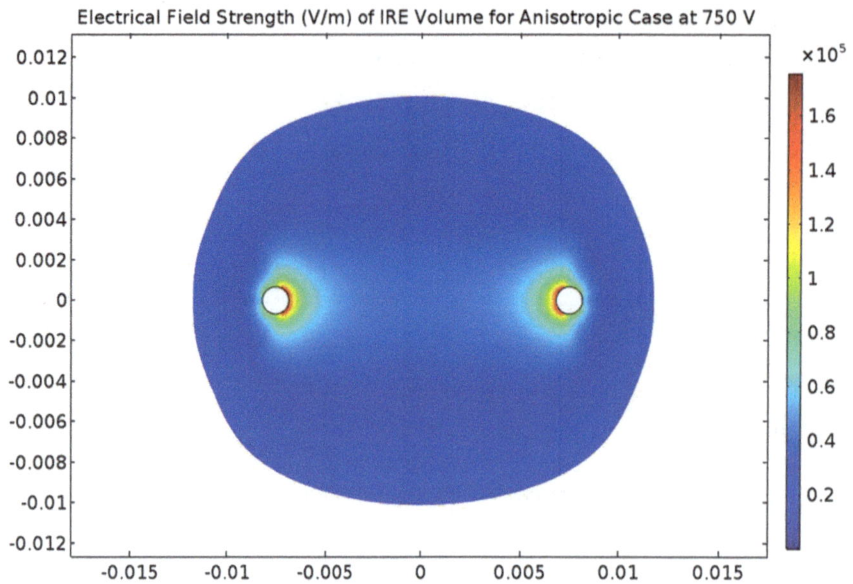

Fig. 5.5 Plot showing the IRE volume for the anisotropic simulation of 750 V. Colors represent electrical field strength. *White regions* correspond to areas where the electrical field was not strong enough to induce cell death from IRE. Notice how the volume is slightly less circular than the isotropic conductivity case seen in Fig. 5.4

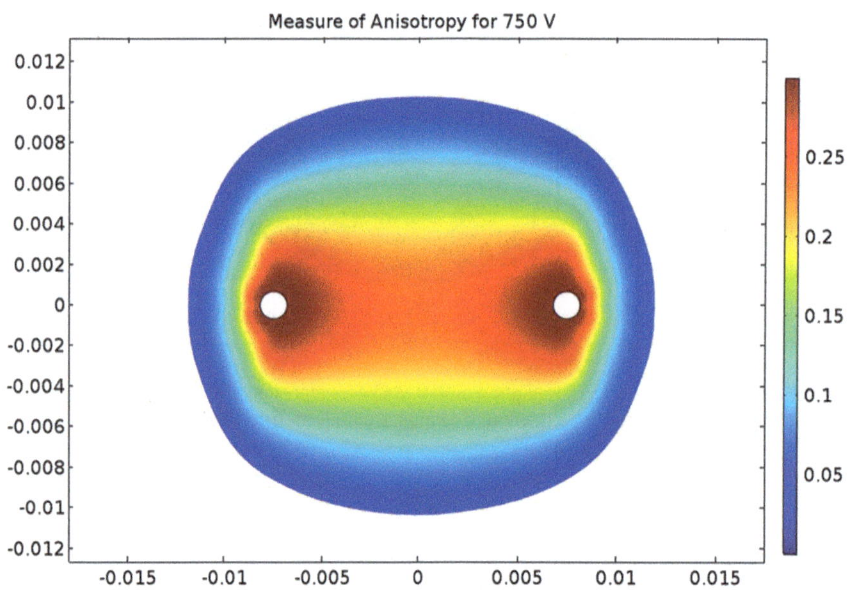

Fig. 5.6 Plot of σ_Δ for the 2D simulation using the results in [1]. The *red regions* correspond to anisotropic conductivity, while the *white areas* correspond to isotropic conductivity

References

1. Castellví, Q., Banús, J., Ivorra, A.: 3D assessment of irreversible electroporation treatments in vegetal models. In: 1st World Congress on Electroporation and Pulsed Electric Fields in Biology, Medicine and Food and Environmental Technologies, pp. 294–297. Springer (2016)
2. Corovic, S., Lackovic, I., Sustaric, P., Sustar, T., Rodic, T., Miklavcic, D.: Modeling of electric field distribution in tissues during electroporation. Biomed. Eng. Online **12**(1), 1 (2013)
3. Dev, S.B., Dhar, D., Krassowska, W.: Electric field of a six-needle array electrode used in drug and dna delivery in vivo: analytical versus numerical solution. IEEE Trans. Biomed. Eng. **50**(11), 1296–1300 (2003)
4. Golberg, A., Rubinsky, B.: A statistical model for multidimensional irreversible electroporation cell death in tissue. Biomed. Eng. Online **9**(1), 1 (2010)

5. Ivorra, A., Al-Sakere, B., Rubinsky, B., Mir, L.M.: In vivo electrical conductivity measurements during and after tumor electroporation: conductivity changes reflect the treatment outcome. Phys. Med. Biol. **54**(19), 5949 (2009)
6. Joshi, R., Schoenbach, K.: Electroporation dynamics in biological cells subjected to ultrafast electrical pulses: a numerical simulation study. Phys. Rev. E **62**(1), 1025 (2000)
7. Jourabchi, N., Beroukhim, K., Tafti, B.A., Kee, S.T., Lee, E.W.: Irreversible electroporation (nanoknife) in cancer treatment. Gastrointest. Interv. **3**(1), 8–18 (2014)
8. Laird, A.K.: Dynamics of tumour growth. Br. J. Cancer **18**(3), 490 (1964)
9. Mezeme, M.E., Kranjc, M., Bajd, F., Serša, I., Brosseau, C., Miklavčič, D.: Assessing how electroporation affects the effective conductivity tensor of biological tissues. Appl. Phys. Lett. **101**(21), 213702 (2012)
10. Miklavčič, D., Šemrov, D., Mekid, H., Mir, L.M.: A validated model of in vivo electric field distribution in tissues for electrochemotherapy and for dna electrotransfer for gene therapy. Biochim. Biophys. Acta (BBA) Gen. Subj. **1523**(1), 73–83 (2000)
11. Neal, R.E. II, Garcia, P.A., Kavnoudias, H., Rosenfeldt, F., Mclean, C.A., Earl, V., Bergman, J., Davalos, R.V., Thomson, K.R.: Simulation of in vivo irreversible electroporation renal ablations. In: 6th European Conference of the International Federation for Medical and Biological Engineering, pp. 813–816. Springer (2015)
12. Pavlin, M., Miklavčič, D.: Theoretical and experimental analysis of conductivity, ion diffusion and molecular transport during cell electroporation – relation between short-lived and long-lived pores. Bioelectrochemistry **74**(1), 38–46 (2008)
13. Weaver, J.C., Chizmadzhev, Y.A.: Theory of electroporation: a review. Bioelectrochem. Bioenerg. **41**(2), 135–160 (1996)
14. Wimmer, T., Srimathveeravalli, G., Gutta, N., Ezell, P.C., Monette, S., Maybody, M., Erinjery, J.P., Durack, J.C., Coleman, J.A., Solomon, S.B.: Planning irreversible electroporation in the porcine kidney: are numerical simulations reliable for predicting empiric ablation outcomes? Cardiovasc. Intervent. Radiol. **38**(1), 182–190 (2015)

Chapter 6
Two-Dimensional Culture Systems to Investigate Mechanical Interactions of the Cell

Brett N. Napiwocki, Alana Stempien, Jacob Notbohm, Randolph S. Ashton, and Wendy Crone

Abstract Creation of well-controlled two dimensional cell culture systems continues to be important for the advancement of basic cell biology investigations as well as innovative platforms for drug development, toxicity testing, and diagnostic assays. These cell culture systems have become more advanced in order to provide the appropriate biomechanical and biochemical cues that better mimic the milieu of *in vivo* conditions. Additionally the platforms themselves can aid in the measurement of cell/cell and cell/substrate mechanical interactions through the use of microscopy techniques. Here we review state-of-the art cell culture systems that involve a range of different substrates, protein micropatterning, and markers for measurement of strains and tractions, including those developed by our labs. We will evaluate the advantages and disadvantages of several culture systems with a focus on those used with stem cell-derived cell types and provide examples of the quantitative measures that can be made with techniques such as traction force microscopy, digital image correlation, and scanning gradient Fourier transform analysis.

Keywords Cell culture systems • Microcontact printing • Traction force microscopy • Digital image correlation • Scanning gradient Fourier transform analysis

There are a multitude of both two dimensional (2D) and three dimensional (3D) cell culture systems that have been reported in the literature, for example [1–6]. Creation of well-controlled 2D cell culture systems continues to be important for the advancement of basic cell biology investigations as well as innovative platforms for drug development, toxicity testing, and diagnostic assays. These cell culture systems have become more advanced in order to provide and to quantify the appropriate biomechanical and biochemical cues that better mimic the milieu of conditions present *in vivo*. Such state-of-the art cell culture systems involve a range of different substrates, protein micropatterning, and markers for measurement of strains and tractions, including those developed by our labs.

The platforms themselves can be designed to aid in the measurement of cell/cell and cell/substrate mechanical interactions through the use of microscopy techniques. When amenable to mechanical measurement, internal forces and displacements within cells and between neighboring cells as well as the forces and displacements that cells induce on their surroundings can be determined.

In much of our work we focus on 2D cell culture systems where micropatterning techniques can be used to create extracellular matrix (ECM) patterns on substrates ranging from glass [7] to polymeric materials such as poly acrylamide (PA) and polydimethylsiloxane (PDMS) with tunable stiffness [8–10]. Our work has also included the use of sacrificial films [8] to transfer ECM patterns to substrates where direct microcontact printing is not feasible and also includes the use of click chemistry, copper-free click reactions [11], to allow cell spreading at a later time point. These systems are also amenable to co-culture with the timed introduction of multiple cell types as illustrated in Fig. 6.1. Importantly, the polymeric substrates utilized can be easily modified to incorporate fluorescent particles to enable techniques such as traction force microscopy to be utilized [10, 12, 13].

The images captured of the cells and their substrates through bright field microscopy and florescent microscopy, can be further analyzed with methods such as edge detection, digital image correlation, and traction force microscopy. Additionally,

B.N. Napiwocki • A. Stempien • R.S. Ashton
Department of Biomedical Engineering, University of Wisconsin – Madison, Engineering Research Building, 1500 Engineering Drive, Madison, WI, 53706, USA

J. Notbohm • W. Crone (✉)
Department of Engineering Physics, University of Wisconsin – Madison, Engineering Research Building, 1500 Engineering Drive, Madison, WI, 53706, USA
e-mail: crone@engr.wisc.edu

P. Zavattieri et al. (eds.), *Mechanics of Biological Systems, Materials and other topics in Experimental and Applied Mechanics, Volume 4*, Conference Proceedings of the Society for Experimental Mechanics Series, DOI 10.1007/978-3-319-63552-1_6

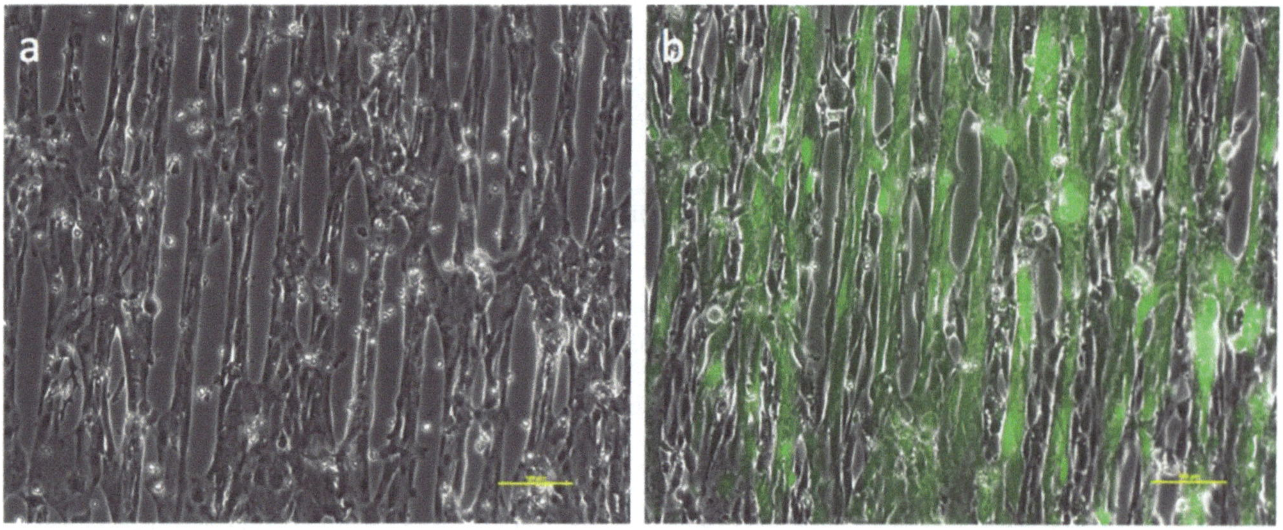

Fig. 6.1 Co-culture of hESC-derived cardiomyocytes (CMs) with mouse 3T3 fibroblasts. hESC-CMs exhibit green fluorescent protein and were seeded within interconnected lanes created my microcontact printing on a 5 kPa PA substrate. (**a**) CMs in lanes prior to 3T3 addition. (**b**) The 3T3 fibroblasts subsequently seeded and cultured for an additional 5 days with green fluorescent protein highlighting CMs. Scale bars 100 μm

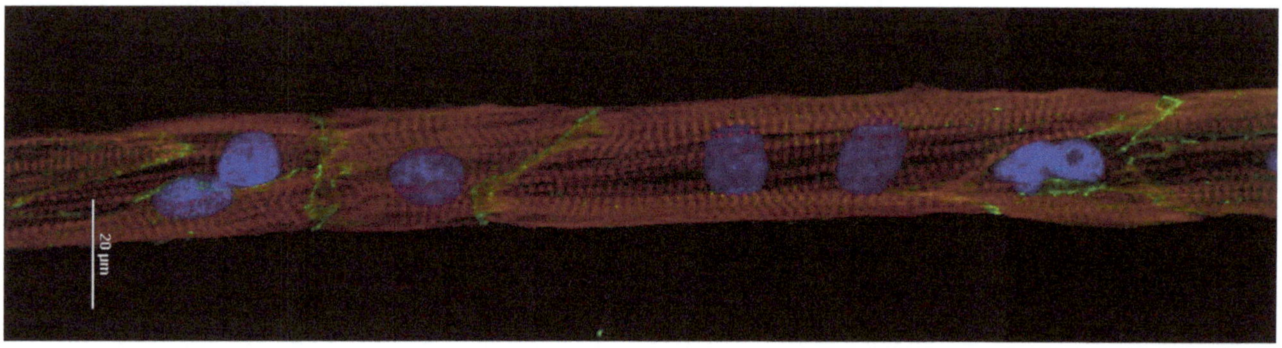

Fig. 6.2 Micropatterned geometry effect on hESC-CMs. Cell alignment and sarcomere organization optimized with lanes of 20-40 μm width. This sample was cultured in Matrigel lanes on a 5 kPa stiffness PDMS substrate for 4 days followed by 14 days of pacing. DAPI (*blue*), actin (*red*), N-Cadherin (*green*). Scale bar 20 μm

cells cultured on a micropatterned substrate can be treated with immunostaining techniques to visualize certain cellular components (such as the ones illustrated in Fig. 6.2 within cardiomyocytes cultured in a micropatterned lane). Techniques such as the Scanning Gradient Fourier Transform (SGFT) method can be used to analyze the thickness, alignment, spacing, and level of organization of the actin within the sarcomere structures of a cardiomyocyte or the radially symmetric N-cadherin structures of a neural rosette, for example [14]. Each substrate system has its own advantages and limitations, making it essential that investigators choose an optimal system for the measurement methods employed. We find that the 2D systems created by microcontact printing techniques are very versatile and provide access to essential data on fundamental behavior of cells. The analysis that can be done with 2D systems can also provide additional measurements not otherwise available with 3D constructs, and are complementary techniques that can be employed in tandem to answer key biology questions. Many 2D constructs allow both single cell and cell aggregate measurements whereas with 3D constructs dissociation of the cells is usually required to make single cell measurements.

References

1. Zhang, B.Y., Xiao, Y., Hsieh, A., Thavandiran, N., Radisic, M.: Micro- and nanotechnology in cardiovascular tissue engineering. Nanotechnology. **22**, 494003 (2011)
2. Almany, L., Seliktar, D.: Biosynthetic hydrogel scaffolds made from fibrinogen and polyethylene glycol for 3D cell cultures. Biomaterials. **26**, 2467–2477 (2005)

3. Ralphe, J.C., de Lange, W.J.: 3D engineered cardiac tissue models of human heart disease: learning more from our mice. Trends Cardiovasc. Med. **23**, 27–32 (2013)
4. Bray, M.A., Sheehy, S.P., Parker, K.K.: Sarcomere alignment is regulated by myocyte shape. Cell Motil. Cytoskeleton. **65**, 641–651 (2008)
5. Ribeiro, A.J., Ang, Y.S., Fu, J.D., Rivas, R.N., Mohamed, T.M., Higgs, G.C., et al.: Contractility of single cardiomyocytes differentiated from pluripotent stem cells depends on physiological shape and substrate stiffness. Proc. Natl. Acad. Sci. U. S. A. **112**, 12705–12710 (2015)
6. Discher, D.E., Mooney, D.J., Zandstra, P.W.: Growth factors, matrices, and forces combine and control stem cells. Science. **324**, 1673–1677 (2009)
7. Salick, M.R., Napiwocki, B.N., Sha, J., Knight, G.T., Chindhy, S.A., Kamp, T.J., et al.: Micropattern width dependent sarcomere development in human ESC-derived cardiomyocytes. Biomaterials. **35**, 4454–4464 (2014)
8. Napiwocki, B.N., Salick, M.R., Ashton, R.S., Crone, W.C.: Controlling hESC-CM cell morphology on patterned substrates over a range of stiffness. In: Mechanics of Biological Systems and Materials, vol. 6, pp. 161–168. Springer, Cham (2017)
9. Napiwocki, B.N., Salick, M.R., Ashton, R.S., Crone, W.C.: Polydimethylsiloxane lanes enhance sarcomere Organization in Human ESC-derived Cardiomyocytes. In: Tekalur, S.A., Zavattieri, P., Korach, C.S. (eds.) Mechanics of Biological Systems and Materials, vol. 6, pp. 105–111. Springer, New York (2016)
10. Notbohm, J., Banerjee, S., Utuje, K.J., Gweon, B., Jang, H., Park, Y., et al.: Cellular contraction and polarization drive collective cellular motion. Biophys. J. **110**, 2729–2738 (2016)
11. Knight, G., Sha, J., Ashton, R.: Micropatterned, clickable culture substrates enable in situ spatiotemporal control of human PSC-derived neural tissue morphology. Chem. Commun. **51**, 5238–5241 (2015)
12. Dembo, M., Wang, Y.L.: Stresses at the cell-to-substrate interface during locomotion of fibroblasts. Biophys. J. **76**, 2307–2316 (1999)
13. Butler, J.P., Tolic-Norrelykke, I.M., Fabry, B., Fredberg, J.J.: Traction fields, moments, and strain energy that cells exert on their surroundings. Am. J. Phys. Cell Phys. **282**, C595–C605 (2002)
14. Salick, M.R., Napiwocki, B.N., Knight, G.T., Ashton, R.S., Crone, W.C.: The scanning gradient Fourier transform (SGFT) method for assessing sarcomere organization, alignment, and heterogeneity, In preparation

Chapter 7
Continuous Composite Beams Stiffened with CFRP Sheet at the Hogging Moment Region

Ayman El-Zohairy, Hani Salim, Suzan Mostafa, and Hesham Shaaban

Abstract The improvement in the flexural response and cracking resistance of steel-concrete composite beams stiffened with Carbon Fiber Reinforced Polymers (CFRP) sheets at the hogging moment regions is investigated experimentally and numerically in this study. The experimental program consists of two composite beams tested in an inverted position to simulate a portion of a continuous composite beam at the hogging moment region. The concrete flanges of the beams are reinforced externally with corrugated steel deck which used also as a permanent formwork. The experimental findings are used to develop and validate a reliable finite element (FE) model which able to simulate the non-linear performance of the stiffened beams as well as plain beams. In addition, a FE analysis on full-scale continuous composite beams with reinforced concrete flange is carried out to confirm the influence of using CFRP strips as a strengthening technique at the hogging moment regions. The beam deformation, steel flange strain, concrete strain, and CFRP strain confirm good agreement between the FE analysis and the test data. The beam capacity is enhanced only by 11% due to a premature failure in the composite flange. Although the composite connection between the concrete flange and the steel beam is improved by enhancing the cracking resistance of the concrete, the ductility of the strengthened beams is reduced by 20% in comparison to the un-strengthened beam.

Keywords Experimental • Finite element • Composite beam • Corrugated steel deck • CFRP sheets

7.1 Introduction

For large spans in bridges and building systems, continuous steel-concrete composite beams have been used to moderate the sagging moment values at the mid-spans. The steel section and reinforcement in the concrete flange play a vital role in determining the beam capacity at the hogging moment regions over the intermediate supports [1]. Construction time can be saved by using the corrugated steel deck as a platform and then as external reinforcement for the concrete flange. The behavior of the continuous beam is improved by adding CFRP sheet to the concrete flange surface at the hogging moment region. This improvement is achieved by controlling possibly existing cracks in the concrete flange under service loads. The higher stiffness, fatigue life, creep rupture, and less susceptibility against aggressive environments are the main advantages of CFRP materials [2].

In the literature, previous works have been recorded on the performance of composite beams with corrugated steel deck. A FE model was proposed and validated using previous experimental findings to simulate the non-linear behavior of concrete, steel beam, steel rebars and the shear connectors [3]. The effect of shear connectors' position and thickness of the corrugated steel deck on the capacity, ductility, and modes of failure for the shear connectors was explored [4]. The composite beams with deeply corrugated steel deck showed higher ductility and capacity [5]. A parametric study on a proposed composite flange for steel girders that spans longer but weighs less than the normal concrete flange was carried out using eight tested beams [6].

In addition, several works have been reported on the performance of composite beams stiffened with CFRP bonded to the steel beam to enhance the beam capacity at the sagging moment regions. Experimental tests were carried out on composite beams to explore the possibility of different strengthening techniques of CFRP strips at the sagging moment

A. El-Zohairy (✉) • H. Salim
Department of Civil and Environmental Engineering, University of Missouri, Columbia, MO, USA
e-mail: aeq76@mail.missouri.edu; SalimH@missouri.edu

S. Mostafa • H. Shaaban
Structural Engineering Department, Zagazig University, Zagazig, 44519, Egypt
e-mail: Suzanaa1@hotmail.com; Hesham_fawzy2000@yahoo.com

P. Zavattieri et al. (eds.), *Mechanics of Biological Systems, Materials and other topics in Experimental and Applied Mechanics, Volume 4*, Conference Proceedings of the Society for Experimental Mechanics Series, DOI 10.1007/978-3-319-63552-1_7

regions [7, 8]. Experimental study on composite timber beams stiffened by steel sections and CFRP was carried out to show an improvement in the beam capacity, deformation, and ductility for the upgraded beams in comparison to the reference beams [9]. FE models were developed and validated using existing experimental work to compute the flexural capacity of composite beams strengthened with CFRP plate and to predict the debonding failures as well as plate-end debonding [10–12]. An experimental program, consists of six continuous composite beams, was carried out to inspect the influence of using CFRP sheet at the hogging moment region in sustaining the composite connection between the steel beam and the concrete flange [13]. The test findings assert the efficiency of CFRP sheets in keeping composite action for continuous composite beams and avoiding crack propagation in the concrete flange under the applied load. A new method for bending upgrading of steel beams using a specific kind of drillable FRP laminates was developed [14]. Furthermore, the possible increasing in the beam capacity stiffened with mechanically anchored CFRP–GFRP laminates was explored by testing 11 continuous steel beams.

Although previous experimental works have been done to explore the performance of steel-concrete composite beams upgraded with CFRP, most of these past works dealt only with CFRP attached to the steel beam at the sagging moment regions or composite beams without corrugated steel deck. Only a few attempts were done on composite beams strengthened with CFRP at the hogging moment regions. Furthermore, the ribs in the corrugated steel deck are oriented perpendicular to the steel beams which leads to difficulty in expecting the structural performance of the composite beam or the shear connectors with this orientation. Therefore, this work aims to set up an experimental work and develop a reliable nonlinear FE model to evaluate the performance of composite beams with corrugated steel deck and stiffened with CFRP sheet at the hogging moment region.

7.2 Experimental Work

Two simply supported beams with composite deck, C1, and C2, are tested in an inverted position to simulate a portion of a continuous composite beam at the hogging moment region. These beams are designed to achieve full shear connection between the composite concrete flange and the steel beam to avoid all types of undesired modes of failure like shear cone failure and shear connector failure.

7.2.1 Details of the Tested Specimens

The typical details of the tested beams are presented in Fig. 7.1. The length of each beam is 2000 mm and tested over a clear span of 1800 mm. The total thickness of the composite concrete flange is 100 mm with 500 mm width. The ribs in the composite deck are oriented perpendicularly to the steel beams to simulate secondary beams in the flooring systems of bridges. Although the corrugated steel deck is used as external reinforcement for the concrete flange, an additional

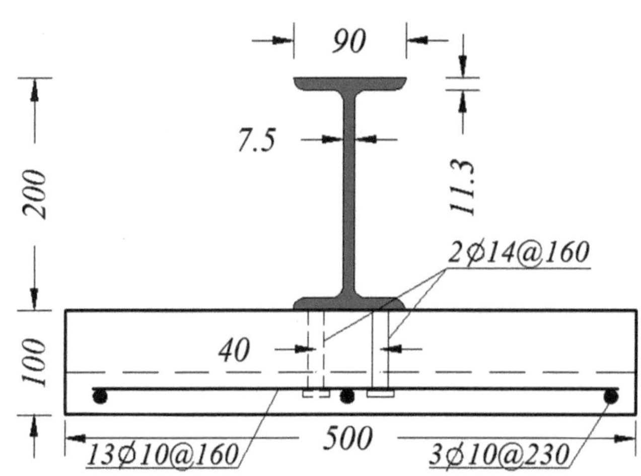

Fig. 7.1 Typical geometry of the tested specimens

reinforcement is added to the concrete flange to avoid any failure due to the longitudinal shear force. One row of shear connectors with 14 mm in diameter and 80 mm in height is welded to the top flange of the steel beam in a staggered arrangement with 200 and 40 mm spacing in the longitudinal and transverse direction, respectively. For CFRP sheet, each strip is cut to the desired size with a length of 1700 mm and a width of 100 mm. The strips are applied to the tension face of the composite concrete flange one by one and cured at the room temperature for 7 days before being tested.

7.2.2 Material Properties

Three concrete cubes of 150 × 150 × 150 mm are poured for compression tests. These cubes and the concrete decks are cured in the laboratory under a plastic sheet. The characteristic concrete strength achieved after 28 days is 29.4 MPa. Three coupons are taken from the flange of the steel beam for compression tests. In addition, tension testing of steel rebars and shear connectors is also completed.

A unidirectional carbon fiber tape, Sikawrap Hex-230C, is used in the strengthening process with an effective thickness of 0.381 mm. The laminate properties provided by the manufacturer are 231 GPa, 1.7%, and 4.1 GPa for the tensile modulus, ultimate strain, and tensile strength, respectively. In addition, a two-component adhesive, Sikadur-330 Normal epoxy, is used to bond the laminates to each other and to the concrete surface. The mixing ratio of the epoxy is 4 to 1 of resin to hardener by weight. The tensile modulus, flexural modulus, ultimate strain, and tensile strength provided by the manufacturer are 4.5 GPa, 3.8 GPa, 0.9%, and 30 MPa, respectively.

7.2.3 Test Set-Up and Instrumentation

Every beam is inverted upside-down with a concrete flange facing down and the loads are applied on the steel flange. This loading technique puts concrete to be exposed to a tensile stress and simulates an existence of concrete flange at the hogging moment region in a continuous composite beam. The typical test set-up is shown in Fig. 7.2.

Accurate strain measurements are obtained by using three strain gauges directly glued to different parts of the composite beam. Strains are measured in the compression flange of the steel beam, the concrete surface at mid-span, and CFRP strips. In addition, one linear variable differential transformer (LVDT) is used to monitor the vertical deformation at mid-span.

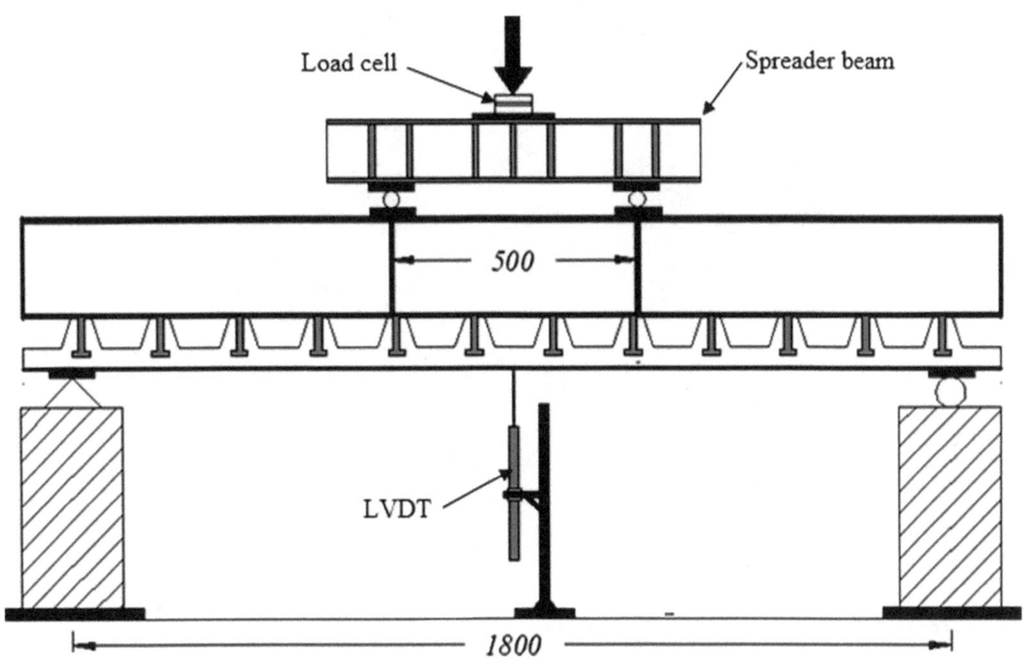

Fig. 7.2 Test set-up

A data logger is used to collect the load value, strains, and deflection from the testing machine, strain gauges, and LVDT, respectively.

7.3 Test Results and Discussion

Table 7.1 summarizes the basic experimental results of the two configurations tested in this study. The table presents the ultimate moment M_u, yield deflection Δ_y, ultimate deflection Δ_u, displacement ductility factor μ, and mode of failure of the tested specimens. Concrete failure followed by yielding in the steel rebars and rib shear failure are the two modes of failure for the reference and strengthened beams, respectively.

7.3.1 Reference Beam (C1)

The first specimen, C1, is used as a reference beam without any retrofit. The specimen is loaded gradually with a loading rate of 350 N/sec. The first crack starts on the surface of the composite concrete flange at mid-span when the moment reaches 15.6 kN·m. Excessive cracks followed by yielding in the steel rebars start at a moment of 75 kN·m. Just prior to the ultimate capacity being reached, diagonal cracks at supports are found accompanied by concrete separation from the corrugated steel deck; as shown in Fig. 7.3. The ultimate capacity of the reference beam is 87.6 kN·m and the ultimate deflection is 11.8 mm.

Table 7.1 Summary of the tested beams results

Beam	M_u (kN·m)	Δ_y (mm)	Δ_u (mm)	μ	Mode of failure
C1	87.6	2.85	11.8	4.1	Concrete failure
C2	97.3	3.9	12.8	3.3	Rib shear failure

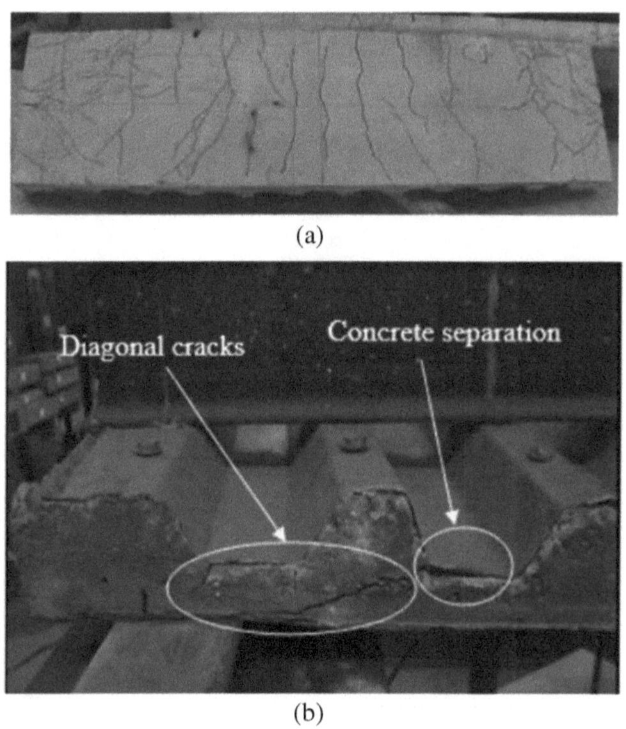

Fig. 7.3 For the reference beam (C1): (**a**) excessive cracks on the concrete surface and (**b**) diagonal cracks at supports

(a)

(b)

Fig. 7.4 For the strengthened beam (C2): (**a**) no failure in CFRP laminates and (**b**) rib shear failure

7.3.2 Beam Retrofitted with CFRP Strips (C2)

The second specimen, C2, is strengthened with one layer of CFRP sheet divided into three strips. This specimen is loaded gradually by the same loading rate as the reference beam. The elastic range of the strengthened beam continues until a moment of 83.1 kN·m and a deflection of 3.9 mm. A premature rib-shear failure occurs without any failure in CFRP strips; as shown in Fig. 7.4. The maximum moment on the strengthened beam is 97.3 kN·m with a maximum deflection of 12.8 mm. The presence of CFRP strips on the concrete surface reduces the propagation of the tensile cracks in comparison to the reference beam.

7.3.3 Effect of Retrofitting on Deflection and Strains

Figure 7.5 presents comparisons between the experimental findings of the tested specimens. Although the ultimate capacity of the strengthened beam is enhanced by 11%, the undesired mode of failure of the composite concrete flange prevents to get the full benefit of using CFRP sheets. Furthermore, a reduction occurs in the ductility of the strengthened beam by 20% which indicates that adding CFRP strips to the concrete surface greatly decreases the deforming capability at the ultimate phase of loading. The positive strain values represent tensile strains in the concrete and CFRP strips. In contrast, the negative strain values indicate compressive strains in the steel flange.

7.4 Finite Element Model

The main purpose of the FE analysis is to develop a numerical model able to simulate the nonlinear flexural performance of the strengthened beams as well as plain beams. Also, the FE analysis is used to extract additional results that have not been monitored during the experimental test. So, the commercial FE package ANSYS [15] is used to implement the FE investigation. The experimental findings are used to validate the FE model.

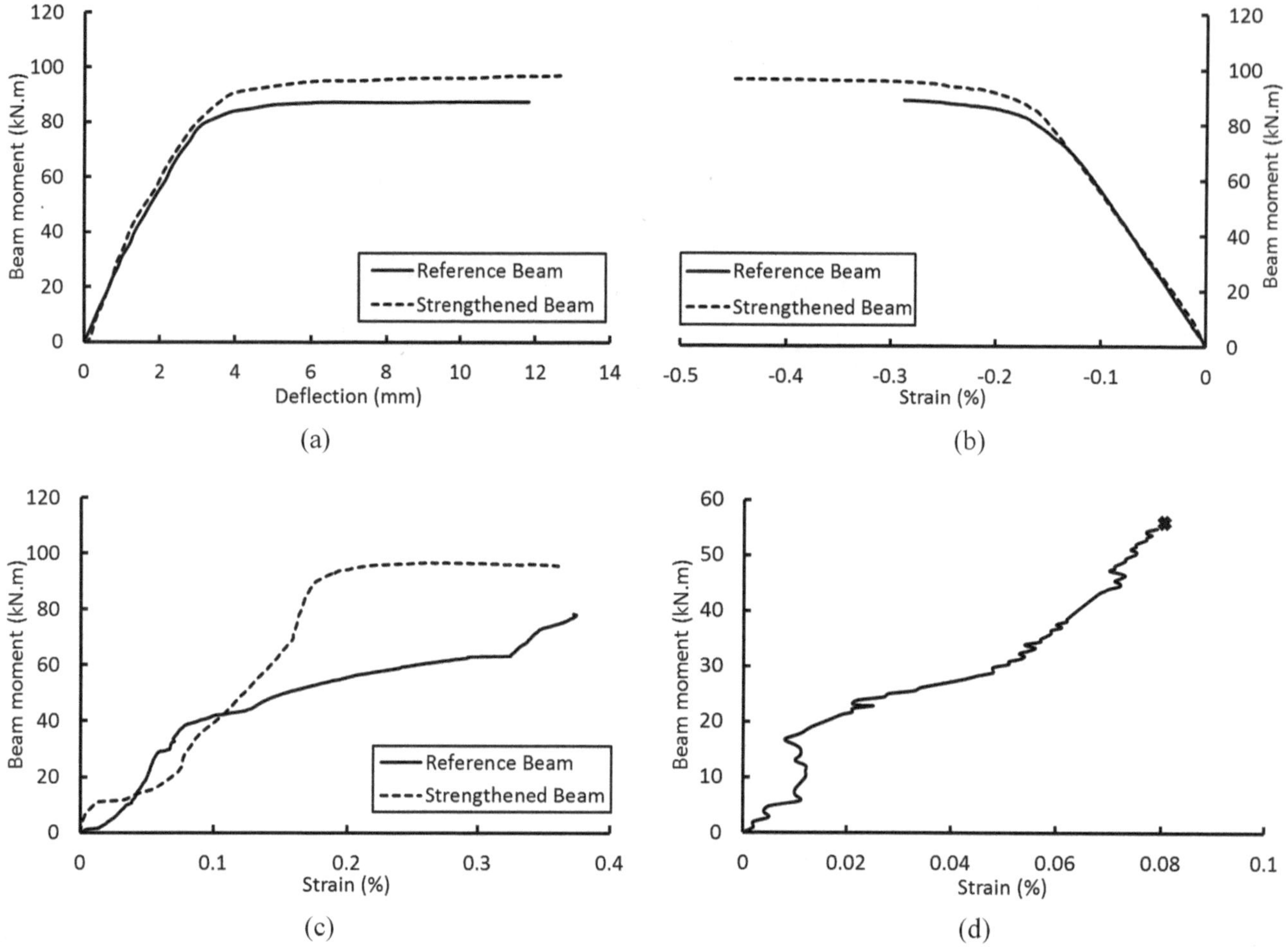

Fig. 7.5 Experimental findings for the tested specimens: (**a**) deflection, (**b**) strain in the steel flange, (**c**) strain in the concrete flange, and (**d**) strain in CFRP laminates

7.4.1 Element Selection

The composite concrete flanges, concrete and corrugated steel deck, are modeled using two elements. The concrete is idealized by using a solid element, SOLID65, which has the capability of cracking in tension and crushing in compression. The corrugated steel deck is represented by four nodes element, SHELL41, with three degrees of freedom at each node. The element has membrane (in-plane) stiffness but no bending (out-of-plane) stiffness. The steel beams, web and flange, are modeled by using a brick eight-node element, SOLID45, which has plasticity, stress stiffening, large deflection, and large strain capabilities. The slippage between the steel beam and the concrete flange induced by the shear stud as a deformable shear connector is simulated using a beam element and a spring element. The uniaxial beam element, BEAM23, is used to simulate the shank of the shear connector. The unidirectional spring element, COMBINE39, is used to resist the longitudinal shear force and simulate the relative movement between the two surfaces in contact. A layered eight-node element, SOLID46, is used to simulate the CFRP laminates. This element can model up to 250 layers with different orientations and orthotropic material properties in each layer. The steel rebars are simulated by using a three-dimensional truss element, LINK8. The interaction between the two surfaces of the steel beam and the concrete flange is simulated by using contact element, CONTAC52, to avoid any penetration and to ensure normal separation between the two surfaces. The normal stiffness is the only real constant that defined for the element and is calculated based on the lowest stiffness of the two surfaces in contact. In addition, a three-dimensional surface-to-surface (contact-pair) element is used to simulate the interface surface between the concrete flange and CFRP laminates. This contact pair assumes full bond between the two surfaces. The mesh geometry, boundary conditions, and loading arrangement of the FE model are presented in Fig. 7.6.

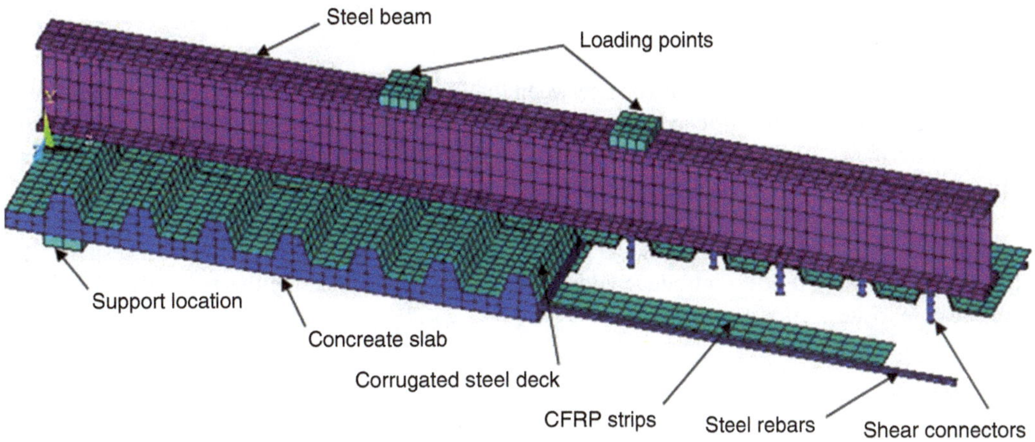

Fig. 7.6 The mesh geometry, boundary conditions, and loading arrangement of the FE model

7.4.2 Material Modeling

The simplified stress-strain relationship for concrete in compression is composed of few points joined together by straight lines. The first point on the curve is a stress (f) equal to $0.3f_c$ and the strain (ε) is computed from Eq. (7.1) using the concrete modulus of elasticity (E_c) [16], while the other points are obtained from Eq. (7.2) [17]. The last point in the relationship is the strain (ε_o) at the maximum concrete strength (f_c) that is computed by Eq. (7.3). Perfectly plastic behavior after the last point on the curve is assumed. On the other hand, the performance of concrete in tension is approximately linear up to the tensile strength of concrete. After this point, the concrete starts to crack and the tensile strength reduces progressively to zero [18].

$$E_c = \frac{f}{\varepsilon} \tag{7.1}$$

$$f = \frac{E_c\,\varepsilon}{1 + \left(\frac{\varepsilon}{\varepsilon_0}\right)^2} \tag{7.2}$$

$$\varepsilon_0 = \frac{2f_c}{E_c} \tag{7.3}$$

In contrast to concrete, bi-linear stress-strain relationships are used to define the uniaxial stress-strain relationships of the steel materials like steel beams, steel rebars, and shear connectors. The steel model is assumed to be an elastic-linear strain hardening material and identical in tension and compression. CFRP laminates are treated as an orthotropic material based on the fiber direction and the properties are nearly the same in any direction perpendicular to the fibers [19].

7.4.3 Real Constants

The relative movement between the steel beam and the concrete flange is taken into consideration by using the load-deformation table in the real constant of the spring element. The analytical relationship that considered for the headed shear connectors is given by Eq. 7.4 [20].

$$P_j = F_{uo}.\left(1 - e^{-0.7\Delta j}\right)^{0.8} \tag{7.4}$$

where: P_j *is* the shear force on the shear connector, F_{uo} is the ultimate capacity of the shear connector, and Δ_j is the slippage between the concrete flange and the steel beam.

7.5 FE Model Validation

To achieve the validity of the FE model, comparisons between the FE results and the test records are illustrated in Figs. 7.7 and 7.8 for the reference beam and the stiffened beam, respectively. These comparisons depend on the beam deformation, strain in the compression flange of the steel beam, strain in concrete, and strain in CFRP strips. Although the FE models show higher capacity over the tested beams, the general behavior of the FE models shows good agreement with the test data. The full bond assumed between the concrete and the steel rebars as well as the corrugated steel deck and the difference in performance of the shear connectors in the FE model are the main reasons for this difference in response at the ultimate stage. The validated FE model is used to introduce additional results to demonstrate the influence of CFRP laminates on different parts of the composite section.

7.5.1 Steel Rebars Stress

For the reference beam, initial tensile cracks appear transversely along the concrete flange at mid-span and abrupt change in stress is observed at cracked moment; as illustrated in Fig. 7.9. Adding CFRP strips to the concrete surface relieves part of the tensile stress in the steel rebars that will be experienced after cracked moment and prevents yielding of the steel rebars to occur when the strengthened beam reaches its ultimate capacity. The same behavior is observed for the steel flange that connected to the concrete flange; as shown in Fig. 7.10.

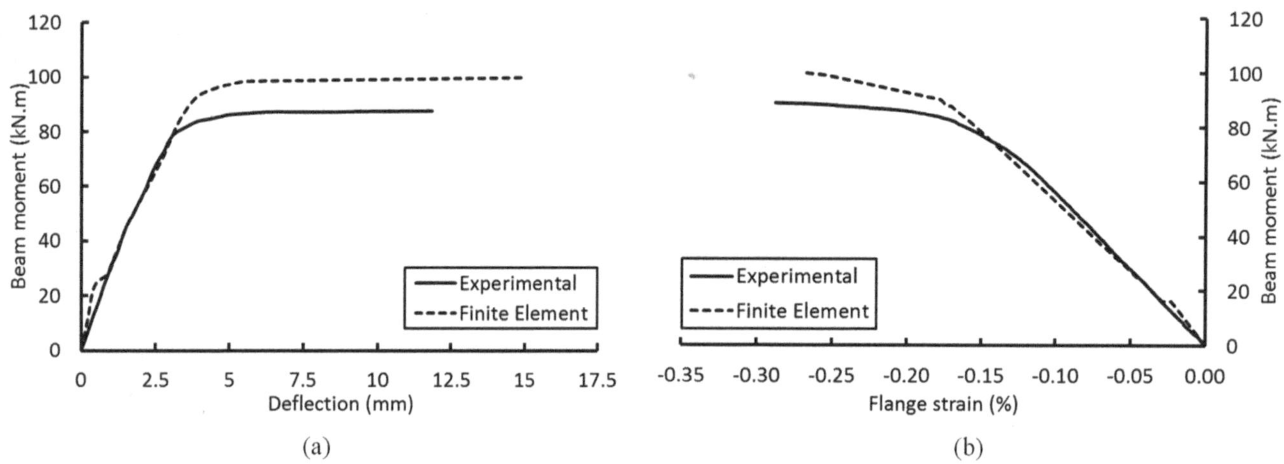

Fig. 7.7 Comparisons between the FE and experimental results for C1: (**a**) deflection, (**b**) strain in the steel flange

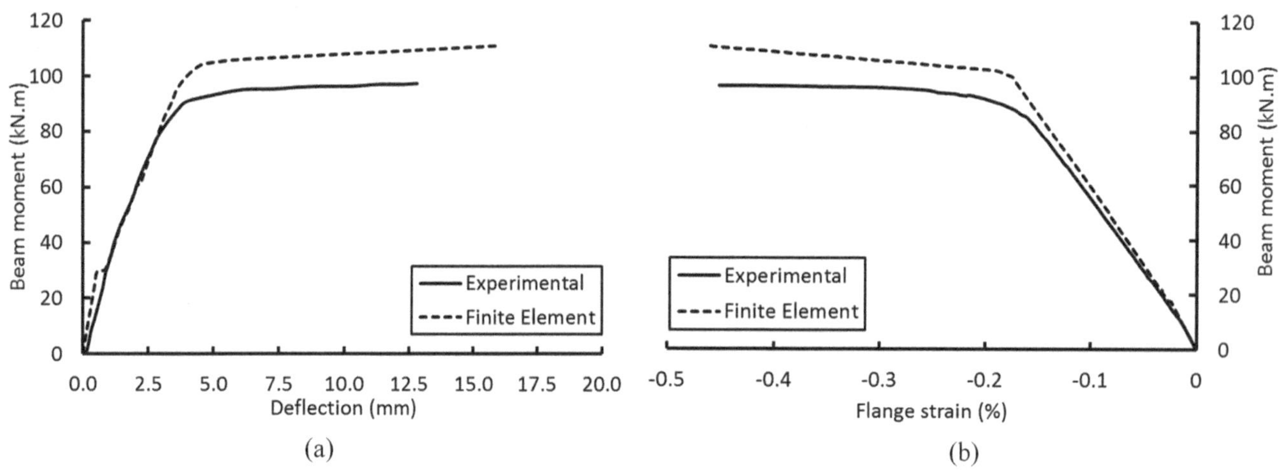

Fig. 7.8 Comparisons between the FE and experimental results for C2: (**a**) deflection, (**b**) strain in the steel flange

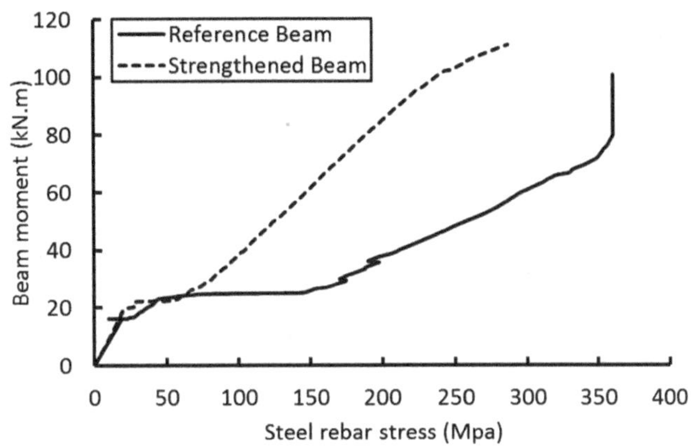

Fig. 7.9 Steel rebars stress

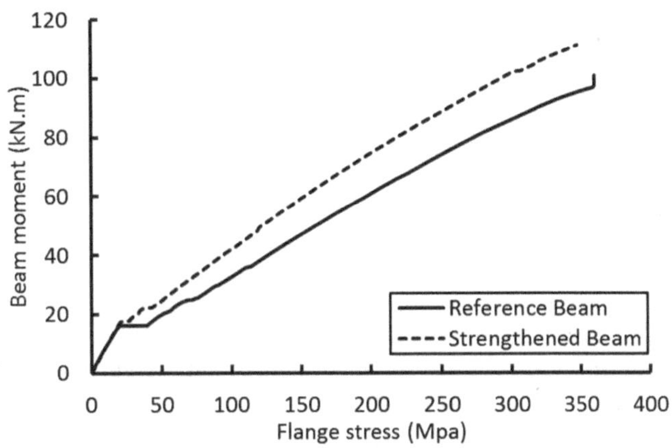

Fig. 7.10 Steel flange stress

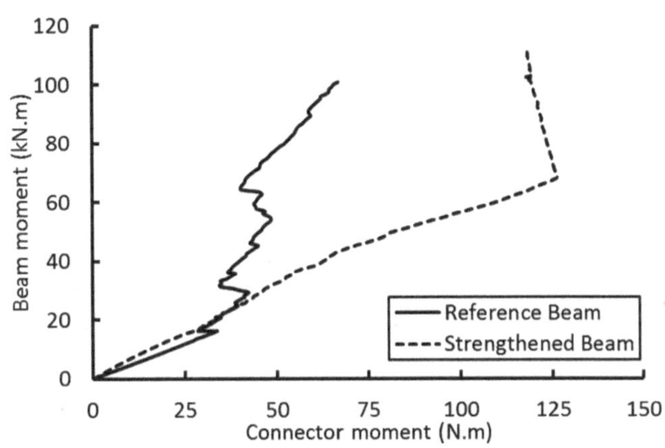

Fig. 7.11 Shear connector moment

7.5.2 Moment Induced in the Shear Connectors

Figure 7.11 illustrates the effect of CFRP laminates on the flexural behavior of the shear connectors located at the beam edge. Before cracked moment, all concrete flanges show the same stiffness and then the same moment induced in the shear connectors. With further loading, a reduction in the concrete flange stiffness occurs due to tensile cracks that formed at mid-span. The presence of CFRP laminates on the concrete surface reduces these tensile cracks and maintains the flange stiffness in comparison to the un-strengthened flange. Because of this drawback, the shear connectors in the strengthened beam resist a moment which is 66% greater than that moment in case of the reference beam. In addition, the slippage between the steel

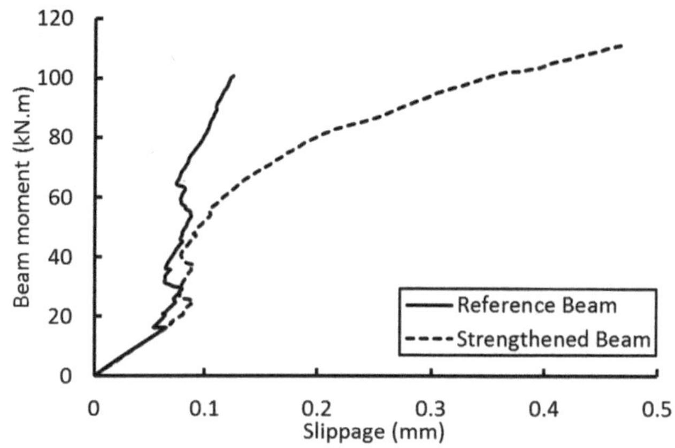

Fig. 7.12 Slippage

beam and the concrete flange is also affected by the flange stiffness. So, this slippage follows the same behavior like that moment induced in the shear connectors; as shown in Fig. 7.12.

7.6 Full-Scale Continuous Composite Beam

The validated FE model is used to build two full-scale continuous composite beams with a reinforced concrete (RC) flange without corrugated steel deck to demonstrate the influence of using CFRP laminates at the hogging moment region. Figure 7.13a, b) illustrates the typical dimensions of two spans of a continuous composite beam. The first beam is used as a reference beam without any retrofit while the second one is strengthened with unidirectional CFRP sheet divided into five strips; as shown in Fig. 7.13c. Each strip is 2200 mm in length, 120 mm in width, and 0.12 mm in thickness. Due to symmetry, a quarter of the full beam with proper boundary conditions is used for modeling.

The analysis results confirm a slight increasing in the strengthened beam stiffness and capacity over the reference beam; as shown in Fig. 7.14. Furthermore, the tensile stresses in the steel rebars are relieved by 93% in comparison to the reference beam; as presented in Fig. 7.15. Yielding in the tension flange of the steel beam at mid-span followed by yielding in the steel rebars at the hogging moment region are the mode of failure for the two analyzed beams.

7.7 Conclusion

This work aims to set up an experimental work and develop a reliable nonlinear FE model to evaluate the performance of composite beams with corrugated steel deck and stiffened with CFRP sheet at the hogging moment region. The main conclusions are:

1. Good agreement between the presented FE model and the test data is obtained which ensured its reliability and accuracy in expecting and analyzing the behavior of composite beams stiffened with CFRP laminates at the hogging moment region as well as plain composite beams.
2. The strengthened beam capacity is enhanced only by 11% due to a premature rib shear failure in the composite concrete flange.
3. Reduction in the beam ductility and limitation on the moment redistribution at the hogging moment regions occur when using CFRP laminates.
4. CFRP laminates succeed to sustain the composite action between the concrete flange and the steel beam by improving the cracking resistance of the concrete flange.
5. The presence of CFRP laminates on the concrete surface relieves part of the tensile stresses in the steel rebars but induces larger moments on the shear connectors in comparison to the reference beam.

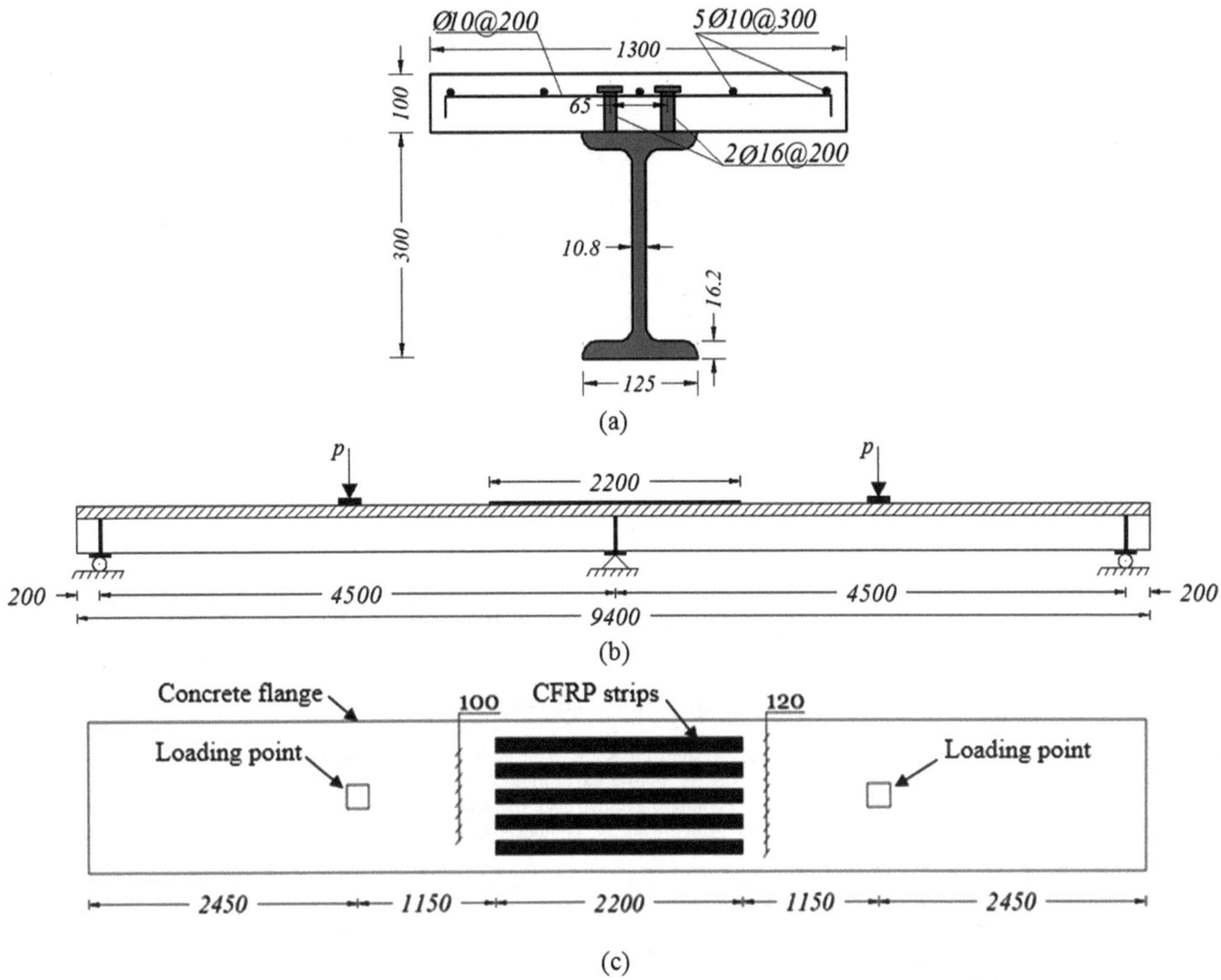

Fig. 7.13 Full-scale continuous composite beam: (a) cross-section, (b) elevation, and (c) arrangement of CFRP laminates

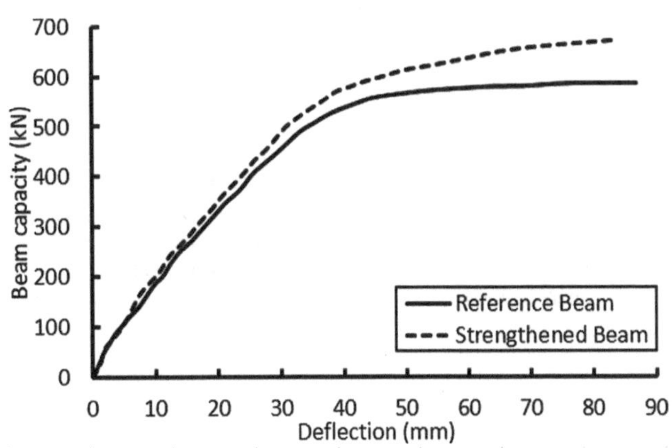

Fig. 7.14 Deflection of the full-scale beam at mid-span

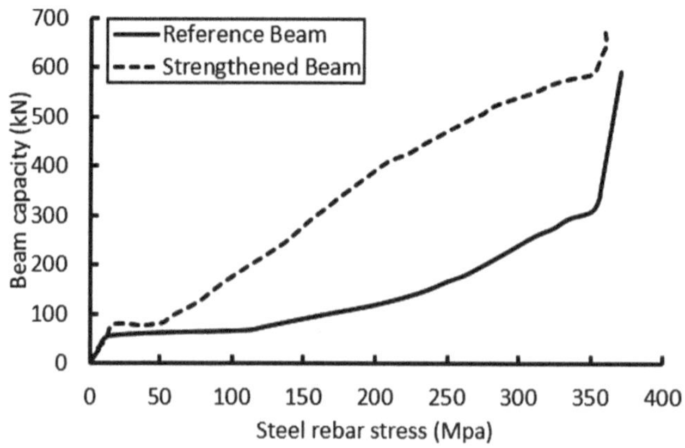

Fig. 7.15 Steel rebar stress at the intermediate support

References

1. AISC, Manual, 13th ed., Chicago (2005)
2. CNR-DT 200, Guide for the design and construction of externally bonded FRP systems for strengthening existing structures, Italy: CNR e Advisory Committee on Technical Recommendations for Construction, p. 154 (2004)
3. Bradford, M. A., Pi, Y., Uy, B.: Ductility of composite beams with trapezoidal composite slabs. In: International Conference on Composite Construction in Steel and Concrete VI (2008)
4. Qureshi, J., Lam, D., Ye, J.: The influence of profiled sheeting thickness and shear connector's position on strength and ductility of headed shear connector. J. Eng. Struct. **33**, 1643–1656 (2011)
5. Ranzi, G., Bradford, M.A., Ansourian, P., Filonov, A., Rasmussen, K.J.R., Hogan, T.J., Uy, B.: Full-scale tests on composite steel-concrete beams with steel trapezoidal decking. J. Constr. Steel Res. **65**, 1490–1506 (2009)
6. Kim, H., Jeong, Y.: Ultimate strength of a steel-concrete composite bridge deck slab with profiled sheeting. J. Eng. Struct. **32**, 534–546 (2010)
7. Tavakkolizadeh, M., Saadatmanesh, H.: Strengthening of steel-concrete composite girders using carbon fiber reinforced polymers sheets. J. Struct. Eng. **129**(1), 30–40 (2003)
8. Schnerch, D., Rizkalla, S.: Flexural strengthening of steel bridges with high modulus CFRP strips. J. Bridge Eng. **13**(2), 192–201 (2008)
9. Ghazijahani, T., Jiao, H., Holloway, D.: Composite timber beams strengthened by steel and CFRP. J. Compos. Constr. (ASCE). **21**(1), 04016059 (2017)
10. Deng, J., Lee, M.M.K., Li, S.: Flexural strength of steel–concrete composite beams reinforced with a prestressed CFRP plate. J. Const. Build. Mater. **25**, 379–384 (2011)
11. Kadhim, M.M.A.: Effect of CFRP plate length strengthening continuous steel beam. J. Const. Build. Mater. **28**, 648–652 (2012)
12. Teng, J.G., Fernando, D., Yu, T.: Finite element modelling of debonding failures in steel beams flexurally strengthened with CFRP laminates. J. Eng. Struct. **86**, 213–224 (2015)
13. Sharif, A., Samaaneh, M., Azad, A., Baluch, M.: Use of CFRP to maintain composite action for continuous steel–concrete composite girders. J. Compos. Const. **20**(4), 04015088 (2016)
14. Sweedan, A.M.I., Alhadid, M.M.A., El-Sawyb, K.M.: Experimental study of the flexural response of steel beams strengthened with anchored hybrid composites. J. Thin-Walled Struct. **99**, 1–11 (2016)
15. ANSYS: Element Manual (10th Ed., Release 4.1). Swanson Analysis Systems, Canonsburg (2008)
16. Gere, J.M., Timoshenko, S.P.: Mechanics of Materials. PWS Publishing Company, Boston (1997)
17. Desayi, P., Krishnan, S.: Equation for the stress-strain curve of concrete. J. ACI. **61**, 345–350 (1964)
18. Bangash, M.Y.H.: Concrete and Concrete Structures: Numerical Modeling and Applications. Elsevier Science Publishers Ltd., London (1989)
19. Kachlakev, D.I.: Finite Element Method (FEM) Modeling for Composite Strengthening/Retrofit of Bridges, Research Project Work Plan, Civil, Construction and Environmental Engineering Department, Oregon State University, Corvallis, Oregon (1998)
20. Slutter, R.G., Ollgaard, J.G., Fisher, J.W.: Shear strength of stud connectors in lightweight and normal weight concrete. J. AISC Eng. **8**, 55–64 (1971)

Chapter 8
Evaluation of Finite Element Codes for the Simulation of Aluminum Extrusion Process

Tushar Bakhtiani, Jing Zhang, and Hazim El-Mounayri

Abstract In order to provide a "standard" and transferable solution to metal extrusion industry, commercial finite element (FE) codes are used to model the highly non-linear deformation process. This paper focuses on evaluating two commercial FEA codes (Deform3D® & HyperXtrude®) that currently represent the state-of-the art of the technology. This research aims at using finite element analysis (FEA) based simulation to study the effect of geometry on the performance of a porthole die for aluminum extrusion, in order to improve the life of tooling and quality of the final part. More specifically, a systematic evaluation approach is followed to compare the two codes. The comparison was based on a thermo-mechanical simulation that predicts: extrusion load, exit velocity, and exit temperature. A comparison of the accuracy of the prediction of each FEA code with the experimental counterparts was used to conduct the relative assessment and select the best code. Other criteria were also used. It was found that HyperXtrude® is an effective finite element simulator that can support the optimization of process parameters such as ram speed, initial billet temperature, initial tooling temperature and bearing length to ensure balanced material flow in order to eliminate product defects. Deform3D®, on the other hand, is an effective tool that not only can be used to assess the quality of extrusion dies by accurately predicting stress in tooling components to adjust critical parameters such as the welding chamber height, mandrel teeth height, web shape and porthole angle in order to improve the die performance, but also be utilized to visualize and achieve a better understanding of the different forming stages during the extrusion process. In comparison to HyperXtrude®, Deform3D® code is not only capable of handling the flow analysis simulations but is also capable of handling tool deflection analysis for complex die geometries with multiple components of different materials, as it allows the user to mimic the real world scenarios of geometric gaps, press fit and die coating. After selecting Deform3D®, the latter was used to create a simulation model for our particular application, and the model was then further validated.

Keywords Extrusion • FEA • Deform3D • HyperXtrude • Die optimization

8.1 Introduction

Extrusion is a complex process which is difficult to solve using analytical techniques. However, using numerical techniques such as finite element analysis helps in formulating a mathematical model that can be solved by a system of simultaneous equations. The metal forming process can be modeled using three different types of finite element formulations: (1) Updated Lagrangian Formulation; (2) Eulerian Formulation; and (3) Arbitrary Lagrangian – Eulerian Formulation. The selection of a specific formulation is determined by two important factors, the problem to be solved and computation resource available.

8.1.1 Updated Lagrangian Formulation

The Updated Lagrangian (UL) formulation is most commonly used in the modeling of extrusion processes, which is divided into numerous time steps that solved sequentially. In the UL formulation the mesh is attached to the deforming material. The advantage of this, is that the transient behavior of material flow such as die filling, weld seam formations and profile exiting effects are accurately simulated, which helps in identifying all the different forming stages occurring during the extrusion

T. Bakhtiani • J. Zhang • H. El-Mounayri (✉)
Department of Mechanical Engineering, Purdue School of Engineering and Technology, Indiana University – Purdue University Indianapolis, 723 West Michigan Street, SL 260, Indianapolis, IN, 46202, USA
e-mail: helmouna@iupui.edu

P. Zavattieri et al. (eds.), *Mechanics of Biological Systems, Materials and other topics in Experimental and Applied Mechanics, Volume 4*, Conference Proceedings of the Society for Experimental Mechanics Series, DOI 10.1007/978-3-319-63552-1_8

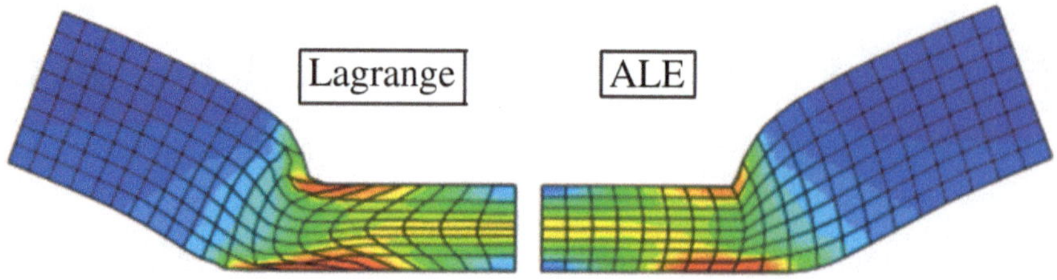

Fig. 8.1 Numerical simulation of coining test reproduced from [1]

process. The only disadvantage of the UL formulation is that large deformation causes the mesh to severely distort, thus leading to multiple remeshes. Remeshing, drastically increases computation time as it creates a new mesh at each iterative step and remaps the information from the old mesh onto the new mesh.

8.1.2 Eulerian Formulation

Eulerian formulation is most commonly used in the field of fluid mechanics but is also suitable for extrusion. In the eulerian formulation the mesh is fixed at a particular location in global coordinate of the system and only spatial variation of state variables is calculated. For utilizing the eulerian formulation the work piece has to be one integral piece that fills the die cavities, which is typically produced through a Boolean technique. The advantage of this method that the mesh does not distort thus saving computational time. However, tracking material flow after it has left the die is difficult to model using the eulerian formulation due to the assumption that the rate of change of strain is zero.

8.1.3 Arbitrary Lagrangia – Eulerian (ALE) Formulation

The ALE formulation is a combination of Lagrangian and Eulerian formulation, as shown in Fig. 8.1. In this formulation the displacement of the material and mesh are decoupled which means that the mesh can move independently of the material reducing mesh distortion which is a major drawback of the Lagrangian formulation and increasing accuracy of the free surface of material flow which is a short coming of the eulerian formulation.

8.2 Simulation Modeling Process

In this section details of aluminum extrusion modelling process used in HyperXtrude® and Deform3D® are presented. The general principles of the numerical modelling process can be categorized into three categories, as shown in Fig. 8.2: (1) Pre-Processing: it is used for creating, assembling or modifying geometries, generating meshes and applying boundary conditions; (2) Simulation Engine: its main purpose is to perform the complex simultaneous numerical calculations and solve for the physical problem being simulated; and (3) Post-Processing: the results generated from the calculations of the simulation engine due to the input parameters of the pre-processing can be represented graphically and exported into data in the post processor.

8.3 Geometric Model

This research aims at using finite element analysis based simulation to study the effect of geometry on the performance of a porthole die for aluminum extrusion, in order to improve the life of tooling and quality of the final part.

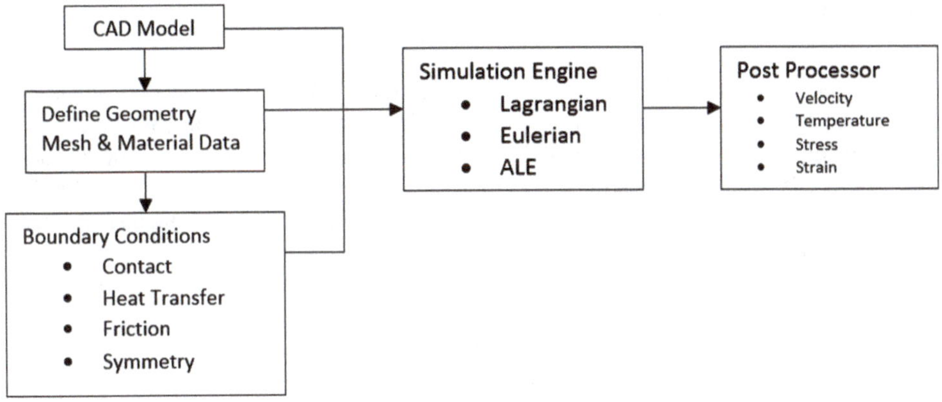

Fig. 8.2 Schematic drawing of a numerical modelling process

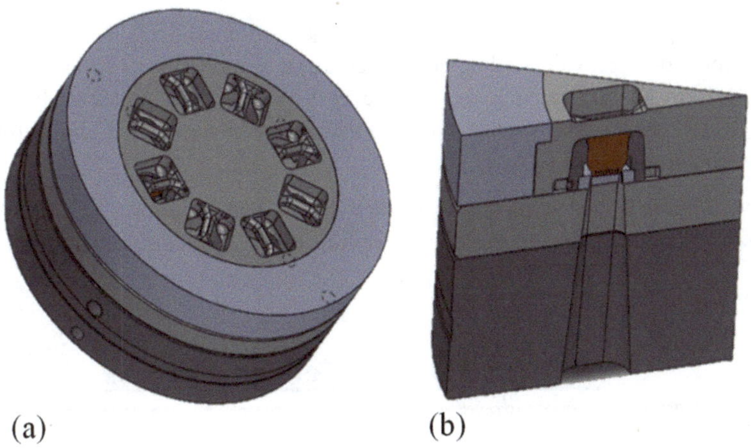

Fig. 8.3 (**a**) Original modular die tool stack, (**b**) 1/16th model

Fig. 8.4 Simulation model of the aluminum extrusion system

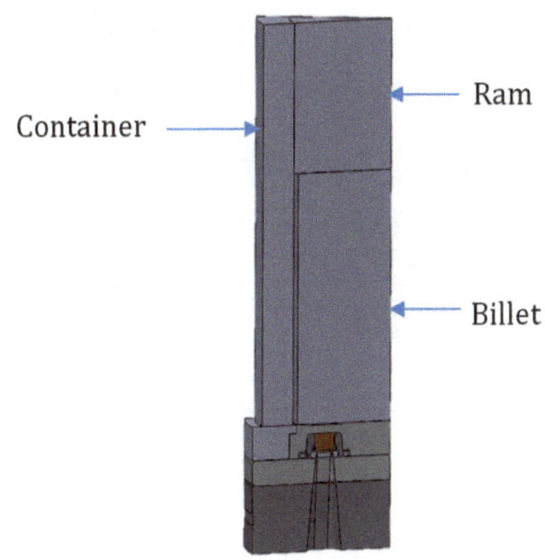

The original modular die tool stack comprises of eight porthole cavities which would take a lot of time to simulate and generate results. Thus, as shown in Fig. 8.3, in order to save computation cost and time the modular die tool stack was reduced to 1/16th of its original geometry due to symmetry.

The simulation model also consists of a ram, container and billet, as shown in Fig. 8.4. Aluminum alloy 3003 series and H-13 tool steel were designated as the material properties of the billet and tooling components.

8.4 Deform3D Simulation Details

Deform3D, developed by Scientific Forming Technologies Corporation (SFTC) is finite element software, that uses an adaptive meshing technique to predict large scale plastic deformation, three dimensional (3D) material flow, heat transfer and die stresses that occur during a complex bulk metal forming process. Deform3D is specifically designed to meet the analysis and optimization needs of the cold, warm and hot manufacturing operations in all industries. Deform3D® has the option to select either Lagrange, Eulerian or ALE formulation depending on type of manufacturing process to be simulated and computational resources available.

A graphical representation of the sequence of steps required for modelling the extrusion process in Defomr3D is shown in Fig. 8.5.

The recommended sequence for modelling an extrusion simulation process is:

1. CAD Model: Using any computer aided design tool extrusion die tooling geometries can be imported into Deforms graphical user interface using an STL file format.
2. Defining Die Geometry:
   (a) Die Tooling Model: During extrusion the billet deforms and is therefore modelled as a plastic object. Depending upon the objective of the simulation study die tooling components can be modelled as either rigid, elastic or plastic objects.
   (b) Mesh Generation: Deform has the capability to use several meshing techniques such as lagrangian, eulerian and combination of arbitrary lagrangian eulerian and coupled eulerian lagrangian. In Deform mesh density can be controlled using two methods:
      (i) System-Defined: This is an automatic mesh generation method in deform that uses a sustem of weights and assigns windows to control the size of elements during the initial mesh generation and subsequent automatic remeshing.

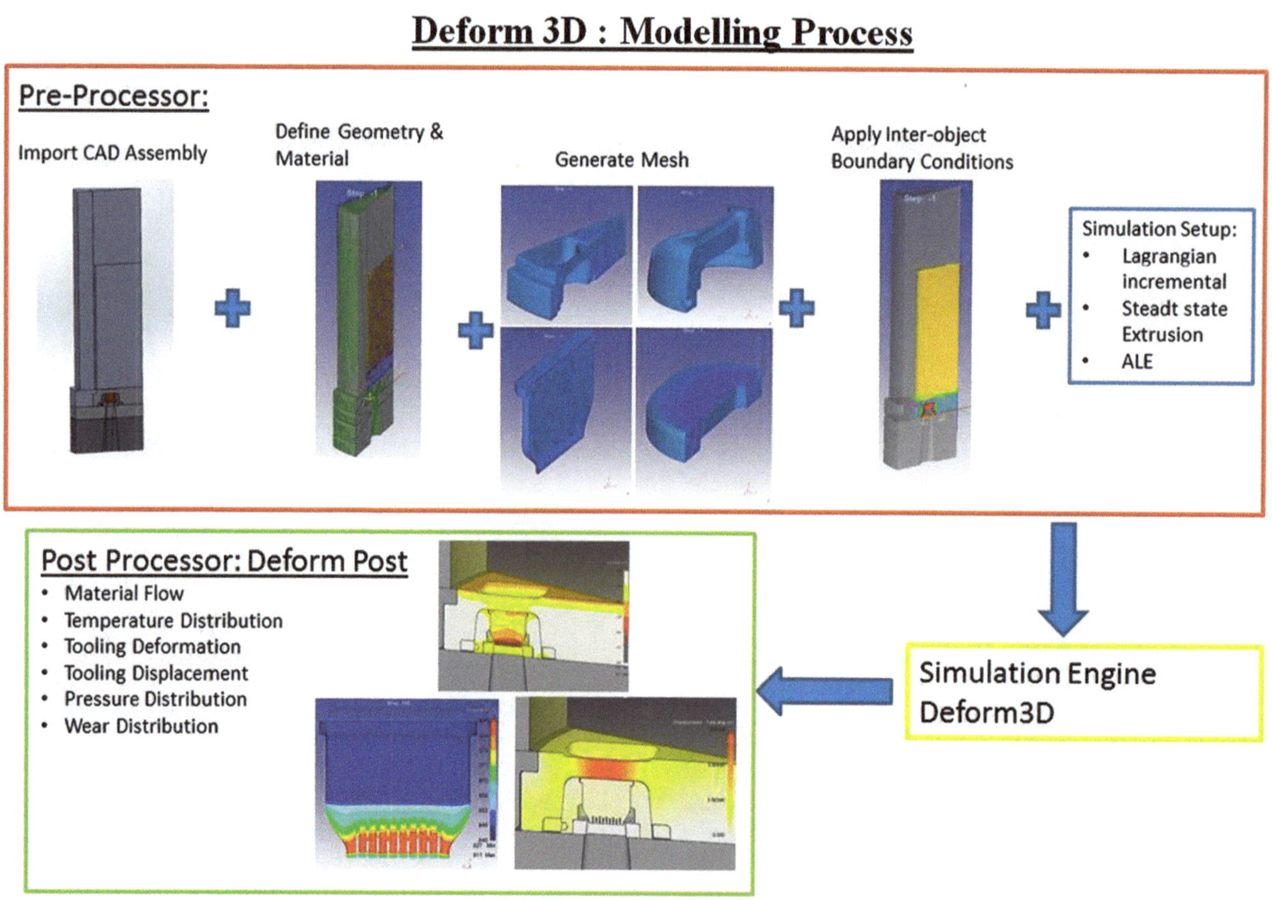

Fig. 8.5 DEFORM3D simulation modelling schematics

(ii) User-Defined: this is a manual mesh generation method that allows the user to specify areas on the object to have higher element sizes relative to other areas of a given object during the initial mesh generation and subsequent automatic remeshing.

(c) Assigning Material Data: Deform has a vast material data library of common extrusion tooling and billet material information, that consist of both mechanical and thermal properties. New material data can also be added if the constitutive equation is known.

3. Boundary Conditions
   (a) Inter-Object relationship: In Deform the purpose of the inter-object realtions is to define how the different die tooling components in a simulation interact with the deforming billet. With the help of inter-object boundary conditions thermal and mechanical behavior of each die tooling component with itself and the billet can be defined. The inter-object relation include the following boundary conditions to be defined:
      (i) Friction Model: The shear friction models available in deform are Shear, Coulomb and Hybrid.
      (ii) Heat transfer Coefficient: specifies the heat transfer between two objects in contact
      (iii) Tool wear: deform has the capability to calculate tool wear of an object which is in contact with another object based on two predefined models used for predicting wear: Archards and Usui.
4. Simulation Settings: in order to get accurate results from the simulation it is important to define the simulations settings such as:
   (a) Type of simulation: Lagrangian incremental, Steady-state machining, Steady-state extrusion, ALE rolling, Ring rolling and ALE extrusion are some of the different simulation types that are available in Deform.
   (b) Solution Method: Conjugate gradient, sparse, GMRES and MUMPS.
   (c) Iteration Method: Direct iteration or Newton-Raphson.
5. Generate simulation database and submit it to solve by the simulation engine.
6. Review the results generated in Deform post processor.

8.5 HyperXtrude Simulation Details

HyperXtrude® is a finite element analysis software that is used to solve complex fluid and heat transfer problems encountered in polymer and metal extrusion processes. It is one of the five engineering performance optimization software packages developed by Altair, that uses the Arbitary Lagrangian-Eulerian (ALE) formulation to predict material flow stresses, material velocity, extrusion load, temperature distribution in tooling and billet, seam weld strength, transverse weld. The ALE method utilized in HyperXtrude®, uncouples the Eulerian (displacement) and Lagrangian (material) mesh thus allowing the mesh to move independently of the material which reduces mesh distortion and increases accuracy [2].

A graphical representation of the sequence of steps required for modelling the flow analysis for extrusion process in HyperXtrude is shown in Fig. 8.6. The recommended sequence modelling an extrusion simulation process in Hyperxtrude is:

1. CAD Model: Using any CAD tool the geometries required for simulation can be imported into HyperXtrudes graphical user interface using a SAT file extension.
   (a) Material flow analysis: only the negative volume of the die cavity needs to be imported to conduct flow analysis.
   (b) Die stress analysis: all the modular die tooling components needs to be combined into a single solid and imported into HyperXtrude to conduct stress analysis.
2. Create Components: The imported negative volume of the billet is split into multiple components to define the volume of the profile, bearing, welding chamber, portholes and the billet, as shown in Fig. 8.7.
3. Extrusion Wizard: it helps the user step by step to input the process parameters
   (a) Specify database directory
   (b) Setup model units: length, velocity, temperature and stress.
   (c) Setup analysis parameters:

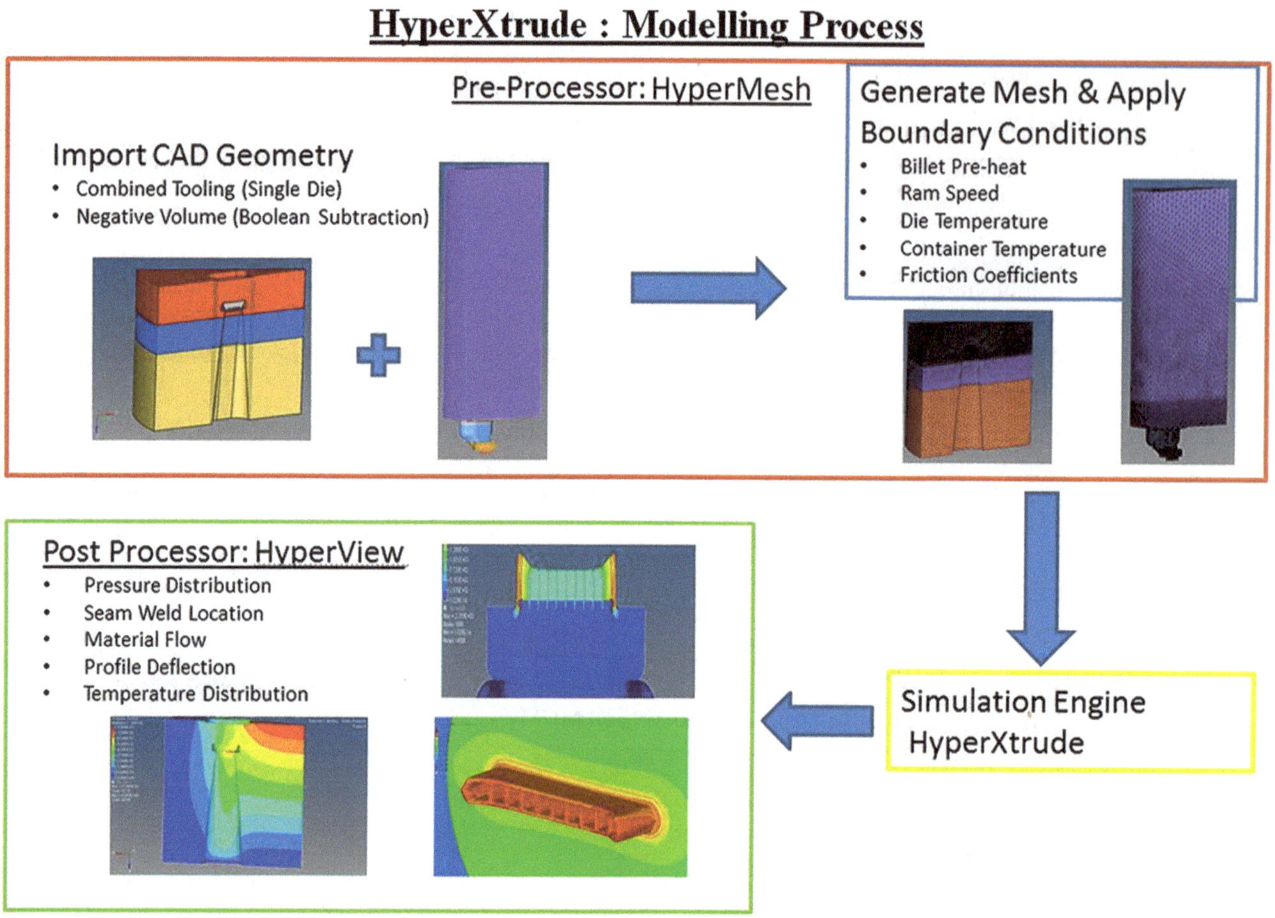

Fig. 8.6 HyperXtrude simulation modelling schematics

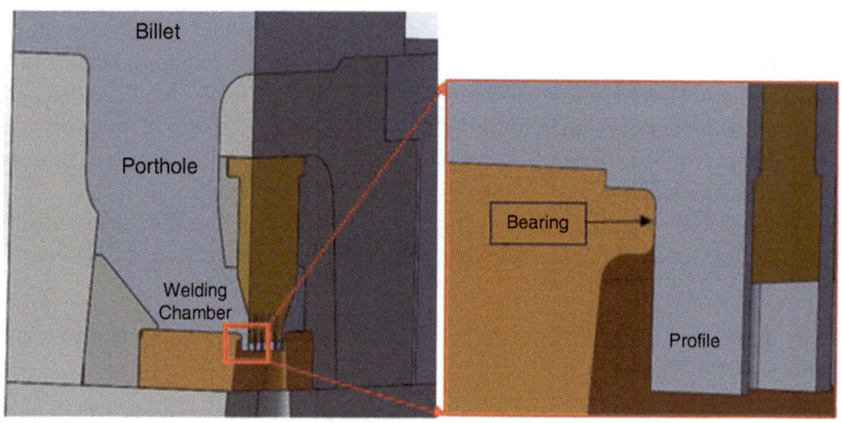

Fig. 8.7 Negative volume components

(i) Profile type: Hollow or Solid
(ii) Extrusion method: Direct or Indirect extrusion
(iii) Analyze profile deformation
(iv) Apply friction conditions between billet and tooling
(v) Billet movement: Fixed or Moving
(vi) Single or Multiple Cycle
(vii) Analyze for skin tracking

(viii) Compute Weld length
(ix) Meshing: Automatic or Manual
(x) Select billet components for meshing

(d) Assign Material Data: HyperXtrude® has a vast material data library for common extrusion tooling and billet material that consist material density, specific heat, thermal expansion ratio, Young's modulus and Poisson's ration
(e) Generate Mesh: HyperXtrude® give the user the ability to either automatically mesh the negative volume or the element size can be specified for manual meshing.
(f) Boundary Conditions:

(i) Billet Pre-heat
(ii) Ram Speed
(iii) Die Temperature
(iv) Container Temperature

4. Define Bearing: this enables HyperXtrude® to optimize the bearing lengths in order to achieve uniform exit velocity of the profile.
5. Save and Export model to radios which is the simulation engine for HyperXtrude®
6. The results of the simulation model can be viewed in HyperView, which is the post processing graphical user interface

To conduct die stress analysis in HyperXtrude, the forces from the flow simulation are interpolated on the combined modular die tooling assembly and the same sequence of steps is followed.

8.6 Comparative Evaluation of Deform3D and HyperXtrue

To evaluate the capabilities and identify the limitations of DEFORM3D® and HyperXtrude® the extrusion simulation of Micro-Multiport tubes through the modular die was conducted. The results obtained from each software is compared in terms of extrusion load, exit velocity, exit temperature, and most importantly die tooling stresses of the modular die [3].

8.6.1 Extrusion Load

Figure 8.8 shows the extrusion load produced through deform simulations. Due to the lagrangian formulation used in deform the entire extrusion load-displacement curve is identified, which shows all the forming stages as the aluminum fills the die cavity.

Figure 8.9 shows the extrusion load obtained from the HyperXtrude® simulations. Due to the use of ALE formulation only the steady state extrusion process is simulated in HyperXtrude® which is observed in the extrusion load.

8.6.2 Exit Velocity

Figure 8.10 shows a samples of the extruded micro-multiport tubes. It is observed that the edges of the tube flow faster than the top and bottom walls.

This characteristic of non-uniform flow was also observed in both simulations, where the exit velocity obtained from HyperXtrud®e and Deform3D® were approximately 38.65in/sec and 45in/sec, respectively (Fig. 8.11).

8.6.3 Exit Temperature

Temperature change during the extrusion process is affected by many factors, including initial billet temperature, initial tooling temperature, extrusion speed, extrusion ratio and friction. However, the maximum temperature is always observed to

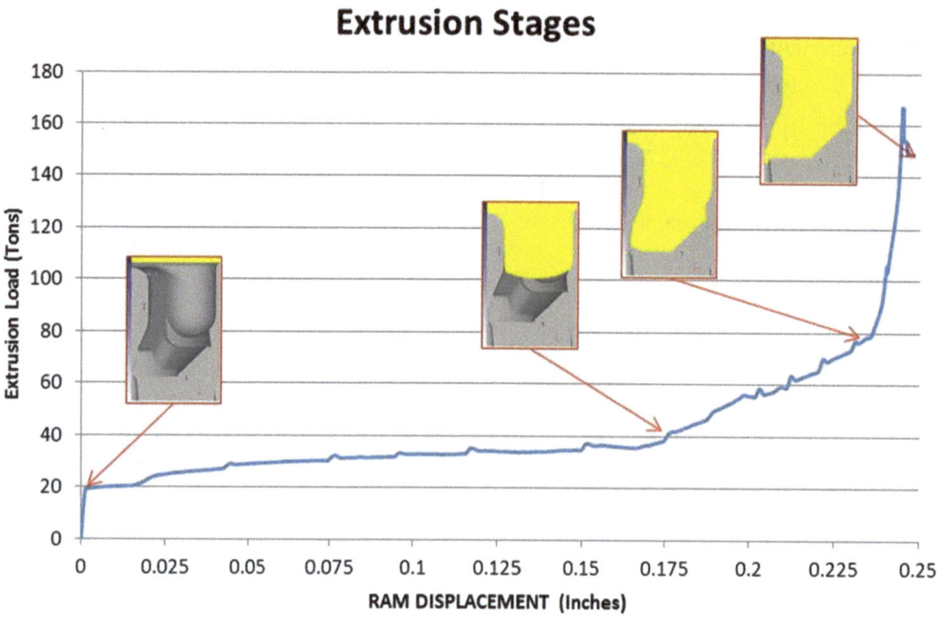

Fig. 8.8 Deform 3D simulated extrusion load

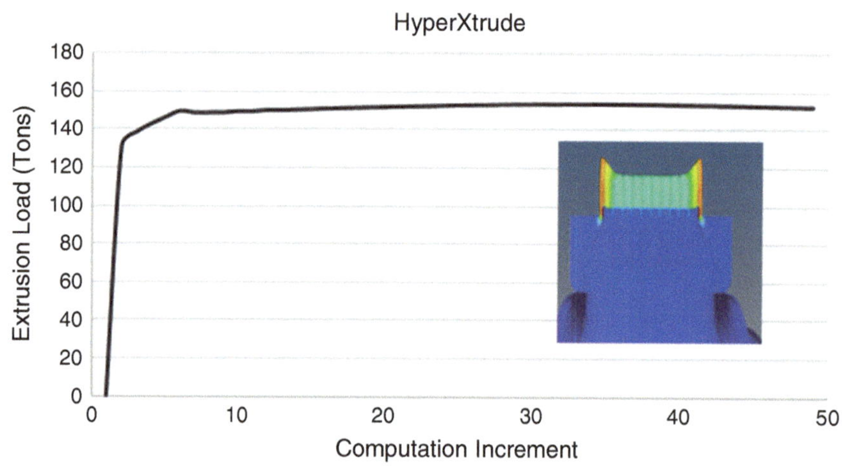

Fig. 8.9 HyperXtrude simulated extrusion load

be close to the bearing region due to the smallest cross sectional area characterized by maximum plastic deformation which generate heat due to loss of energy shown in Fig. 8.12 [4].

Figure 8.13 shows the exit temperature of the Micro-Multiport tube obtained from HyperXtrude® and DEFORM®. It is observed that maximum temperature of the extruding profile starts in the bearing region and extends till the outlet of the profile. The values obtained from HyperXtrude (1070 °F) and Deform3D(1000 °F).

8.6.4 Die Stress

Figure 8.14 shows the stress distribution produced in DEFORM® for each individual components of the modular die. This is very useful information for a die designer or a die manufacturer because it shows the areas where the design improvement is needed before the model is manufactured and used for extrusion.

Figure 8.15 shows the stress distribution produced in HyperXtrude® for the combined modular die tooling geometry. In order to conduct die stress analysis in HyperXtrude® a Boolean addition of all the different components is needed. The

Fig. 8.10 Micro-multiport tube sample

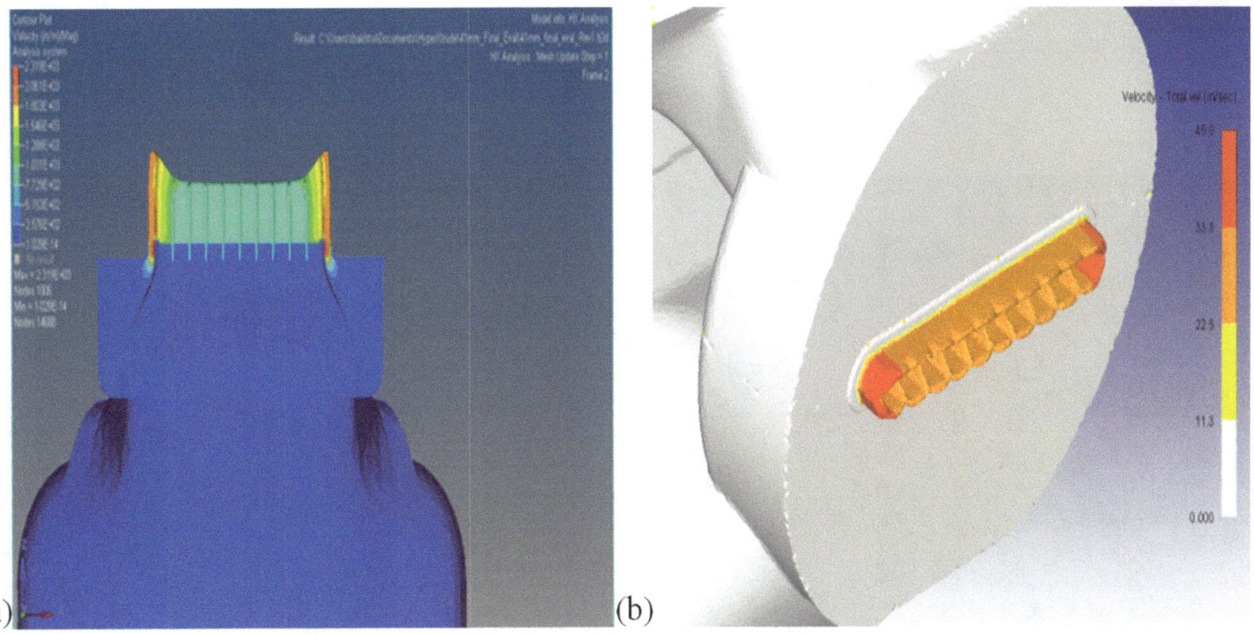

Fig. 8.11 Exit velocity profile (**a**) HyperXtrude, (**b**) DEFORM

effective stress values obtained in HyperXtrude (4.191E+12 psi) were significantly larger than the predicted values of Deform3D® (178,000 psi). Comparing these results to the yield strength of H-13 tool steel (120,000 psi [5]) at elevated temperatures shows that the values obtained from HyperXtrude® are far from the yield strength of H-13 tool steel as compared to the simulated results of DEFORM.

8.7 Conclusion

The aluminum extrusion process is an attractive metal forming process in the manufacturing industry due to its ability to produce quality products with minimal energy and secondary manufacturing operations. Currently, in the industry, the development of complex aluminum extrusion dies are solely based on the experience of the die designer which is continuously corrected after a number of trail extrusion runs before going into production. This trial and error method makes it difficult to guarantee product quality, increase productivity, or achieve cost effectiveness. Utilization of numerical simulations such as HyperXtrude® and Deform3D®, play an important role in replacing the decade old trial and error

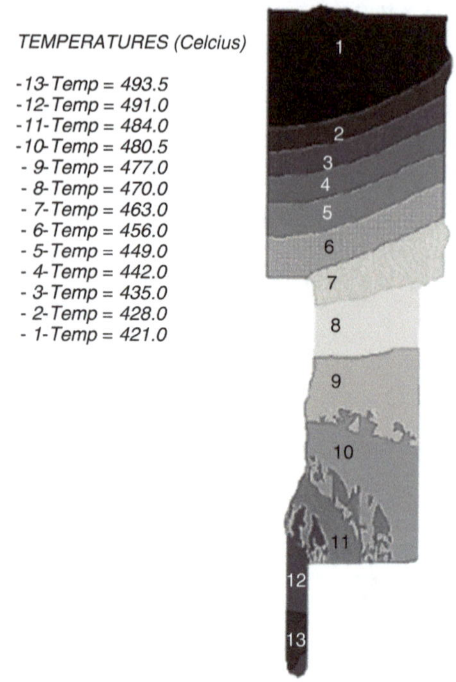

Fig. 8.12 Exit temperature of a hollow profile cross section reproduced from [4]

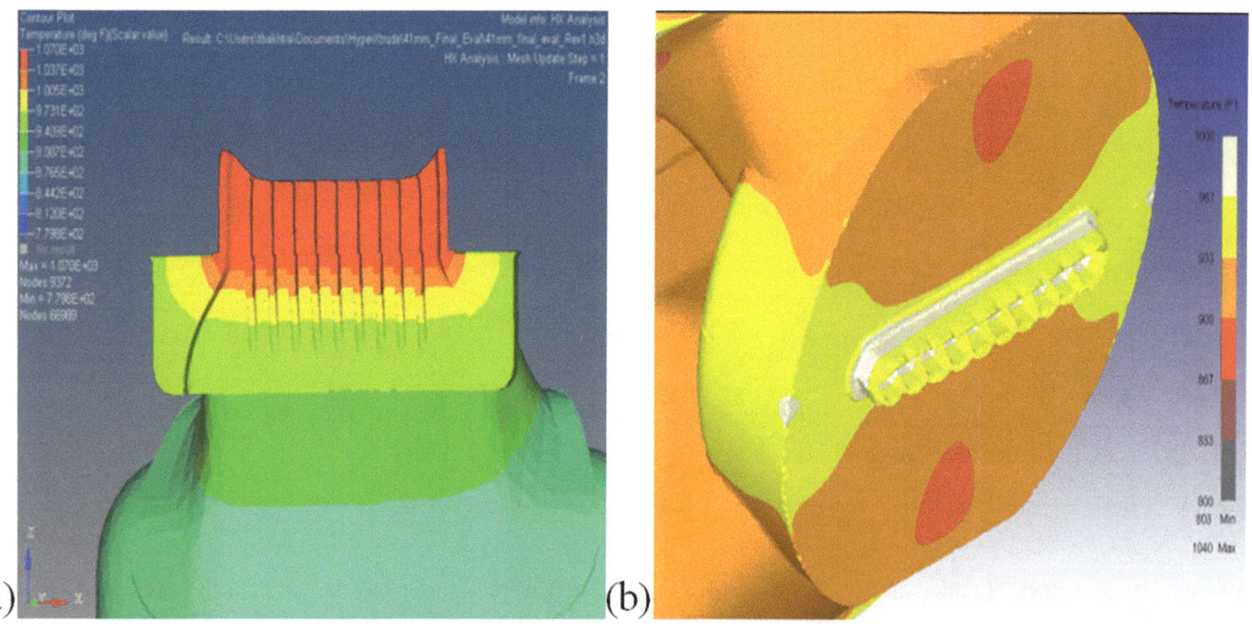

Fig. 8.13 Exit temperature profile (**a**) HyperXtrude & (**b**) DEFORM

approach by empowering extruders to extrude complex profiles with minimal defects and provides the die designer with accurate and theoretical guidance for design improvements at minimal cost.

HyperXtrude® is an effective finite element simulator that helps the extruder in optimization of process parameters such as ram speed, initial billet temperature, initial tooling temperature and bearing length to ensure balanced material flow in order to eliminate product defects thus reducing production costs by eliminating die trails. Deform3D®; on the other hand, is an effective tool that can be utilized by die manufacturers to assess the quality of their die by studying stress of tooling components to make appropriate die improvements such as increasing or decreasing the height of the welding chamber, mandrel teeth height, web shape and porthole angle with accurate guidance. In addition, it can be utilized by extruders to visualize and understand the different forming stages during the extrusion process.

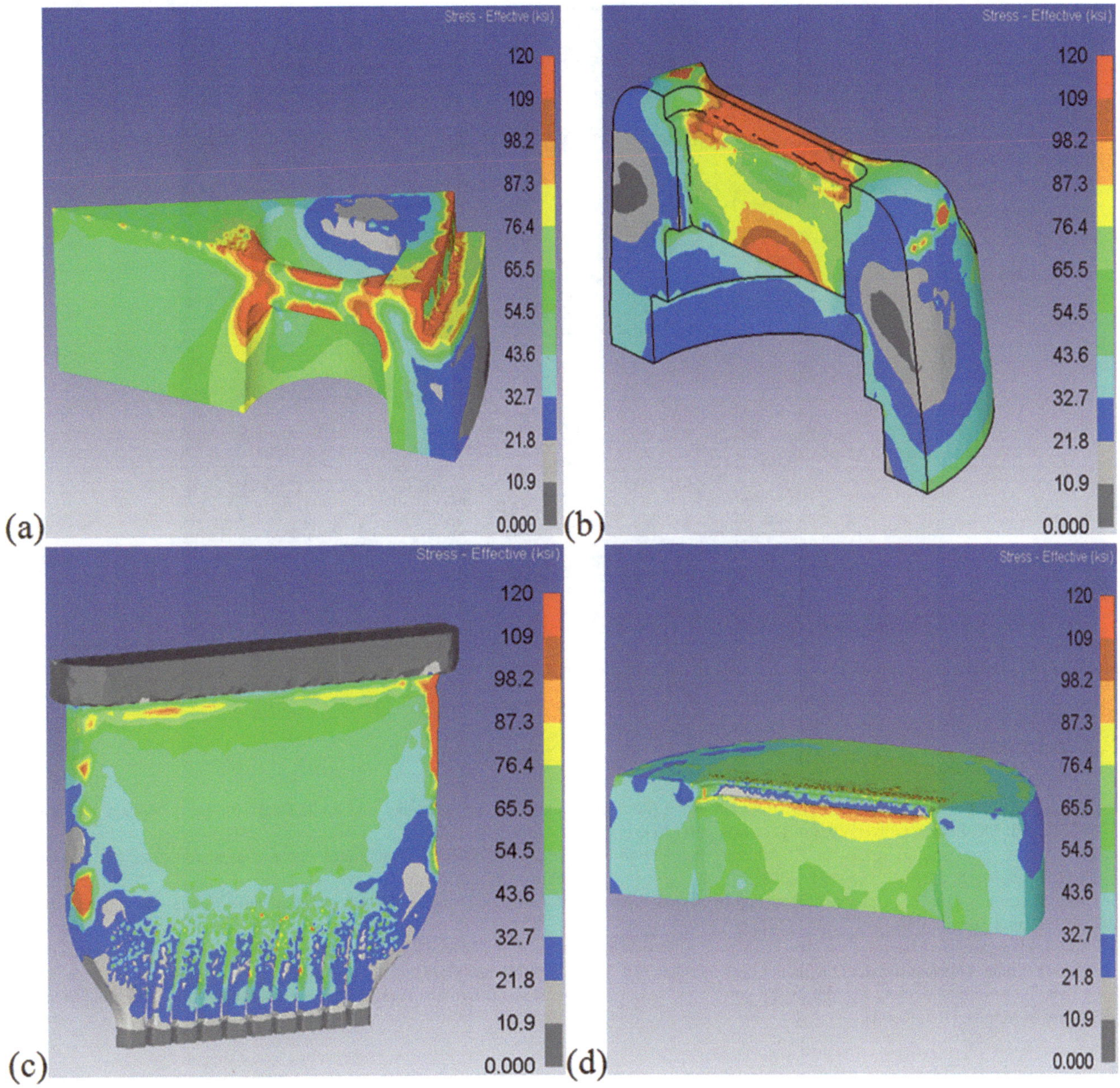

Fig. 8.14 DEFORM simulation stress distribution in modular die tooling components; (**a**) Holder; (**b**) Die Cartridge; (**c**) Mandrel & (**d**) Die Plate

Based on our study, Deform3D® is an appropriate software for aluminum extrusion simulations, as it is not only capable of handling the flow analysis simulations which is vital information for the extrusion process, but it is also capable of handling tool deflection analysis for complex die geometry with multiple components of different materials. As such, it allows the user to mimic the real world scenarios of geometric gaps, press fit, and die coating, among other capabilities.

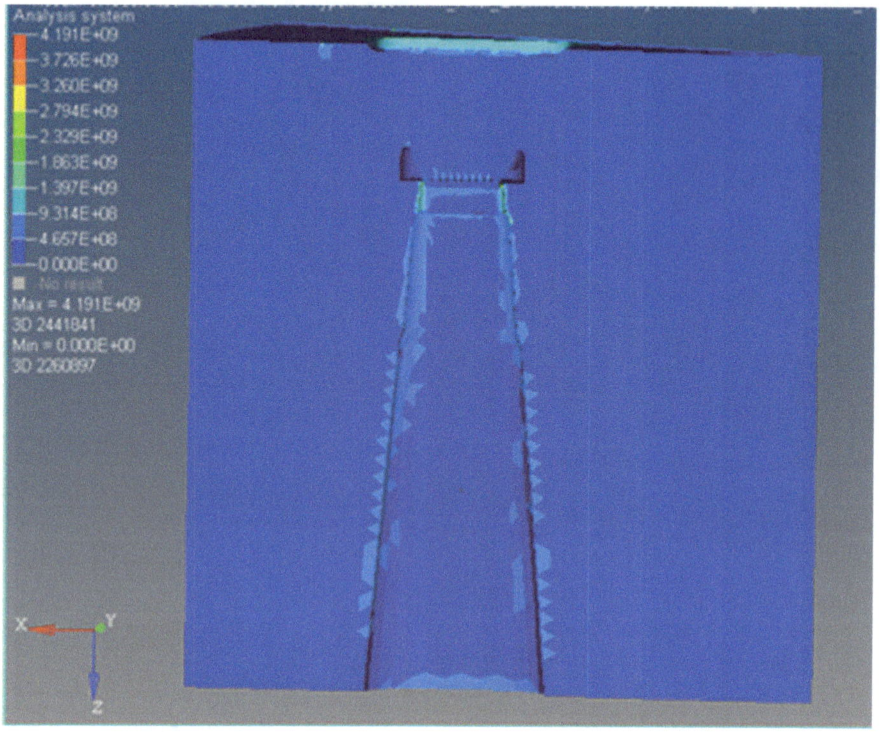

Fig. 8.15 HyperXtrude simulation stress distribution of modular die tool stack

References

1. Linder, D.-I.C.: An Arbitrary Lagrangian-Eulerian Finite Element Formulation for Dynamics and Finite Strain Plasticity Models. Department of Structural Mechanics, University Stuttgart, Stuttgart (2003)
2. Savidis, S.A., Aubram, D., Rackwitz, F.: Arbitrary Lagrangian-Eulerian finite element formulation for geotechnical construction processes. J. Theor. Appl. Mech. **38**(1–2), 165–194 (2008)
3. Tushar Bakhtiani, H.E.-M., Zhang, J.: Modeling of extrusion process of a condenser tube for investigating the effects of mandrel geometry. Int. J. Adv. Manuf. Technol. **92**, 3237–3252 (2016)
4. Flitta, I. Sheppard, T.: Material Flow and Prediction of Extrusion Pressure When Extruding Through Bridge Dies Using FEM. In: Proceedings of International Aluminum Extrusion Technology Seminar, vol. 1, pp. 141–148 (2000)
5. Lyman, T., American Society for Metals: Properties and selection of metals. In: Lyman, T. (ed.) Metals Handbook, vol. 1, 8th edn. American Society of Metals, Ohio (1961)

Chapter 9
The Extended Abstract: The Growth and Mechanical Property of Mesolayer in Abalone Nacre

Bart Prorok, Anqi Zhang, Yan Chen, and Fucito Kristin

Abstract In the red abalone's shell, a special structure which has the unique morphology is named "mesolayer". Its mechanical property is concerned in this research. The mesolayer is built with three sublayers: the columnar-like layer, the organic layer, and the prismatic layer. The three sublayers are tested with nanoindentation test individually and the results are analyzed. And, the nanoindentations on each layer are characterized by SEM.

Keywords Red abalone • Mesolayer • Nanoindentation test

With the highly order hierarchy structure produced through biomineralization process, abalone nacre has long been under investigation. Countless aragonite (one type of calcium carbonate) tablets with 300–500 nm thickness aligned tidily in columns and they are connected with each other by organic thin film (Fig. 9.1). This nacre, which is also called "mother of pearl", shows a 1000 times higher toughness than the mineral aragonite [1].

Beside the tablets structure mentioned above, another special structure, which is named as "mesolayer", appearing repeatedly, especially in the shell of the abalones live in the natural sea environment. Based on the morphology shown by SEM image (Fig. 9.2), the mesolayer could be dived into three more layers: the prismatic layer, the organic layer, and the columnar layer. Both of the prismatic layer and columnar layer are calcium carbonate and some literature indicated that the columnar layer is calcite [2, 3]. The prismatic layer always head forward to the growth direction. All the mesolayer are shown in this three-sublayer structure.

The mesolayer is investigated to show the influence that it plays on the whole nacre hierarchy structure. However, as it is a very thin layer, sometimes it is hard to have mechanical test on each sublayers of the mesolayer accurately.

Parts of our research is to investigate the mechanical property for each individual layer in the mesolayer. Nano-indentation test is used for this part of research. The elastic modulus of each layer was tested and the SEM images for the indentations on each area were taken [4–5]. For the tests applied on the organic area in mesolayer (Fig. 9.3), a flat and stable modulus plot was produced showing that the modulus for it is around 10GPa (Fig. 9.4).

Based on the process of preparing the sample (Fig. 9.5), thickness of the tested organic layer is not quite even and the thickness is hard to be decided.

In some of the tests, when the nano-indentation displacement is deep enough, the influence brought by the substrate started to be shown in the plot (Fig. 9.6).

B. Prorok (✉) • A. Zhang • Y. Chen • F. Kristin
Auburn University, Auburn, AL, USA
e-mail: azz0012@auburn.edu

P. Zavattieri et al. (eds.), *Mechanics of Biological Systems, Materials and other topics in Experimental and Applied Mechanics, Volume 4*, Conference Proceedings of the Society for Experimental Mechanics Series, DOI 10.1007/978-3-319-63552-1_9

Fig. 9.1 The Secondary electron SEM image of California red abalone nacre structure

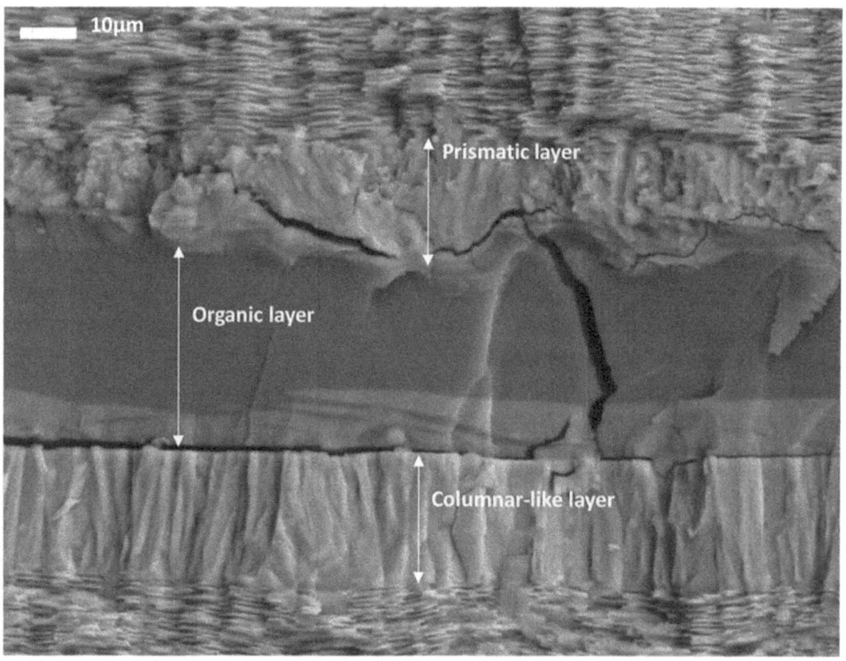

Fig. 9.2 The Backscattered electron SEM image of one mesolayer in a wild California red abalone nacre

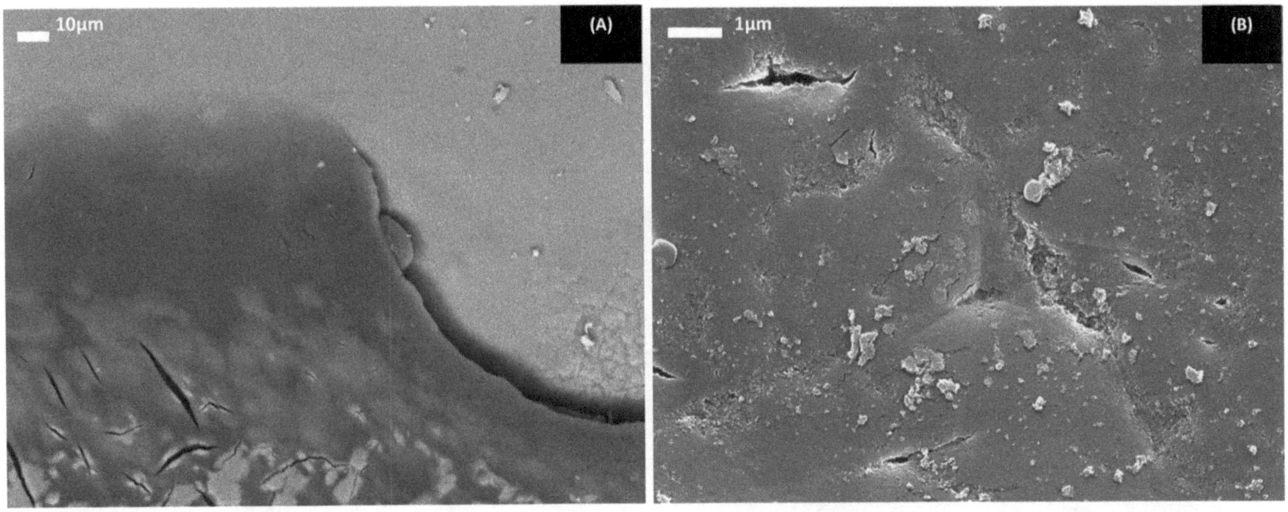

Fig. 9.3 SEM image of the indentation on the organic area of the mesolayer. (**a**) Backscattered electron SEM image shows the contrast (**b**) one of the indentations on the organic area of mesolayer

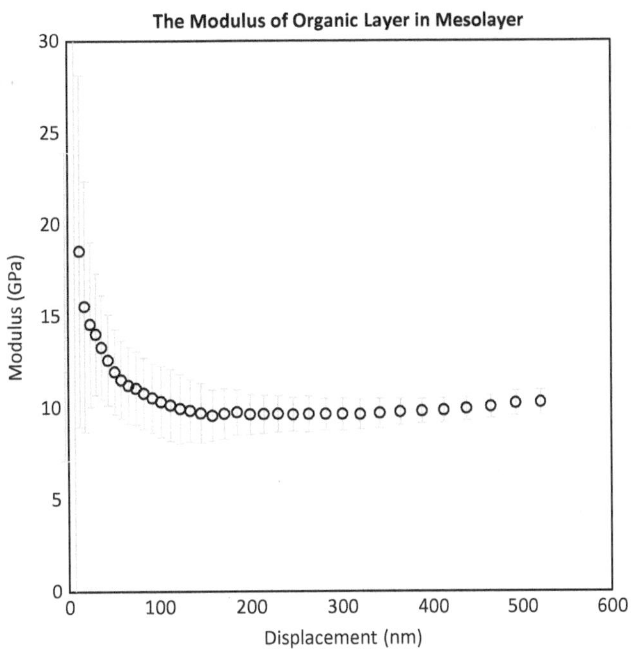

Fig. 9.4 The Nano-indenter tested elastic modulus of the organic area of mesolayer

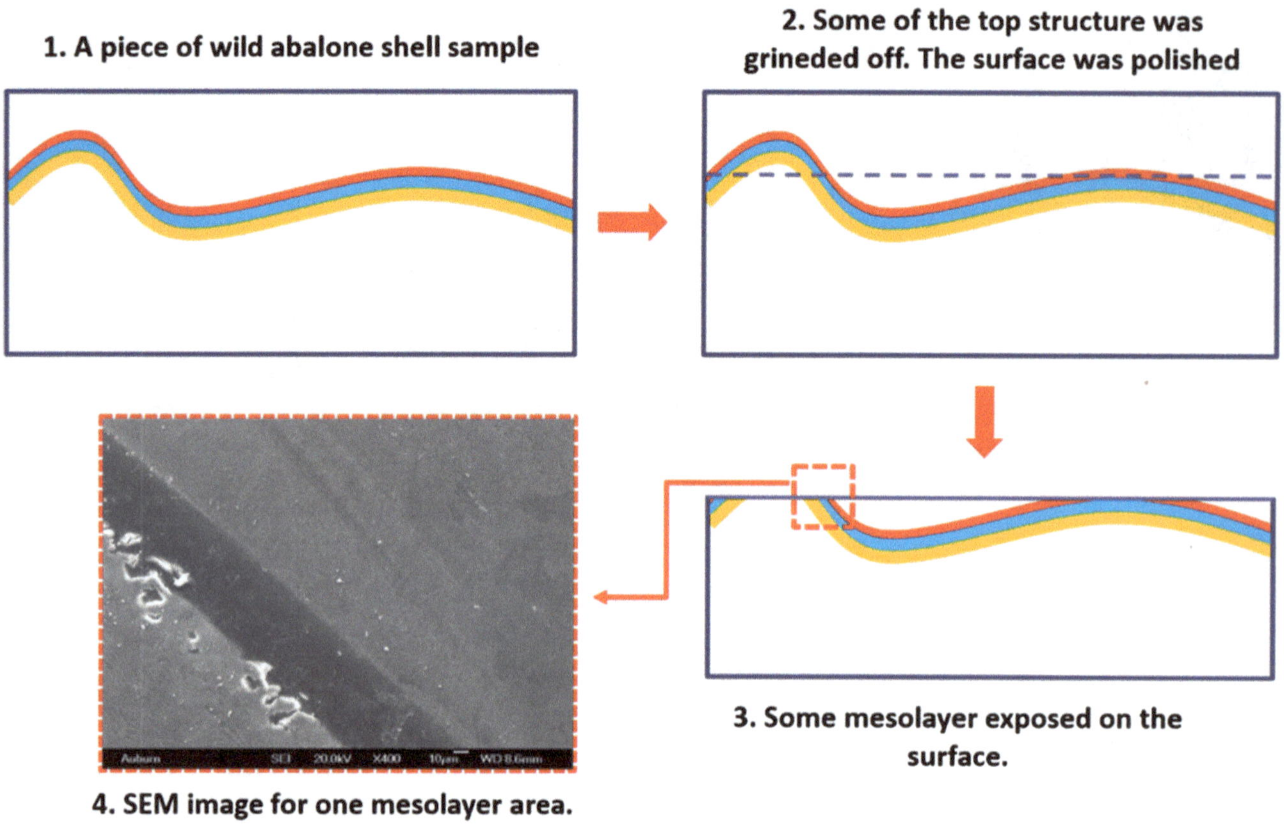

Fig. 9.5 The mesolayer sample preparation process

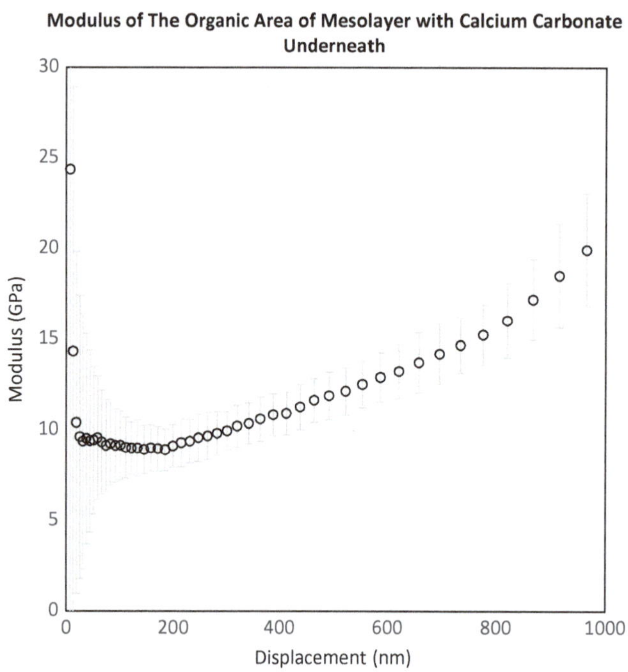

Fig. 9.6 The modulus of the organic area of mesolayer with calcium carbonate composition underneath tested by Nano-indenter

References

1. Barthelat, F., Espinosa, H.D.: An experimental investigation of deformation and fracture of nacre–mother of pearl. Exp. Mech. **47**(3), 311–324 (2007)
2. Blank, S., Arnoldi, M., Khoshnavaz, S., Treccani, L., Kuntz, M., Mann, K., et al.: The nacre protein perlucin nucleates growth of calcium carbonate crystals. J. Microsc. **212**(3), 280–291 (2003)
3. Lin, A.Y.M., Chen, P.Y., Meyers, M.A.: The growth of nacre in the abalone shell. Acta Biomater. **4**(1), 131–138 (2008)
4. Chen, Y., Sullivan, M., Zhang, A., Prorok, B.C.: A new method to extract elastic modulus of brittle materials from Berkovich indentation. J. Eur. Ceram. Soc. (2017). https://doi.org/10.1016/j.jeurceramsoc.2017.08.029
5. Zhou, B., Prorok, B.C.: A discontinuous elastic interface transfer model of thin film nanoindentation. Exp. Mech. **50**(6), 793–801 (2010)

Chapter 10
Dynamic Analysis of a Critical Unmanned Aerial Vehicle Components for Development of Future Structural Health Monitoring Techniques

Michael Coatney, Natasha Bradley, and Jin Yoo

Abstract The aim of this paper is to examine the free vibration characteristics and to understand the natural frequencies of critical components of an Unmanned Aerial Vehicle (UAV). Since microscopic damages such as microcracks, dislocation distribution, crazing, and shear banding in composite and metallic structures can develop into macroscale damage, our goal is to capture the vibration modes and correlate the modes to the damage mechanics. For this study, modal impact tests were performed on an aircraft wing and the vibration mode shapes were captured via Spectral Dynamics, Inc. Bobcat structural analysis system along with a modal impact hammer to create the impact loads. The mode shapes can reveal critical areas where damage has the most probability to occur due to fatigue. We plan to provide detection of these early damage modes as well as predicting the Remaining Useful Life (RUL) of the component.

Keywords Blade • Vibration analysis • Modal analysis • Helicopter • Damage modes • Modal response

Modal Analysis of a stationary structure is the important to understand the dynamic loading conditions. Damage manifests itself in the change in the dynamic structural response, however it is difficult to determine minor damage from the change in the stiffness matrix and does not provide the extent of damage to the members. The change in the stiffness properties or an asymmetric stiffness matrix could be contributed to the varieties in geometry and materials properties of the structure which could lead to complex eigenfrequencies. For this experiment, the modal analysis is compared with a dynamic finite element models used to predict the potential structural instabilities or damage that may occur.

The critical component used is a wing from an RQ-2 UAV. The wing was suspended by way of a custom bracket that makes the wing a cantilever beam. Figure 10.2 shows the 13″ × 6.5″ grid that was created along the air foil an impact hammer was the method used to impact excitation and accelerometers were used to sense the response. Two different methods were used to investigate the bending and torsional mode shapes throughout the air foil. The first method used is a fixed response method. This means that the accelerometer was fixed at a point on the grid, and the impact location will change. Shown in Figs. 10.1 and 10.2, the accelerometer is stationary on the grid, wherein every point represents impact locations throughout the wing section. The second method is the fixed excitation method. This is where the impacts will be created at the same location and the accelerometer(s) will be moved to different grid points on the air foil. Shown in Figs. 10.1 and 10.3, is an image and schematic of a 2D representation of the air foil. It shows the modal hammer impact location, point number 16. The blue squares are simply representing the accelerometer placements. The accelerometers were moved around to all 32 grid points.

In order to create the impacts for the modal analysis, a PCB® Piezotronics Inc. general purpose modal analysis impact hammer was used, and PCB® Piezotronics Inc. ceramic shear ICP® accelerometers were used to sense the responses of the air foil. The BOBCAT 7 by Spectral Dynamics was used to record all excitations and responses, and uses STAR 7 and CATS (Computer – Aided Test – Suite) modal. STAR7 is a complete modal analysis software suitable for small, large, simple or complex structures. STAR 7 creates these frequency response plots per impact location. For example, Fig. 10.4 shows a frequency response of a fixed excitation test. Where the impact location was applied at point 16, and the accelerometers were placed at point 12 and 13. All of the excitations and responses were collected in the Z – Direction. For the initial test, we were unable to catch the natural frequency peaks because the impact frequency were too low at certain impact locations. Although, we could not see all the complex eigenfrequencies, however, a key features was picked up on the distal end of the wing which is where the aileron is located as shown in Fig. 10.5. Also, in the case of the fixed response modal experiments,

M. Coatney (✉) • N. Bradley
US Army Research Laboratory, Aberdeen Proving Ground, Aberdeen, MD, 21005, USA
e-mail: michael.d.coatney.civ@mail.mil

J. Yoo
US Naval Surface Warfare Center (NSWC) Carderock Division, West Bethesda, MD, 20817, USA

P. Zavattieri et al. (eds.), *Mechanics of Biological Systems, Materials and other topics in Experimental and Applied Mechanics, Volume 4*, Conference Proceedings of the Society for Experimental Mechanics Series, DOI 10.1007/978-3-319-63552-1_10

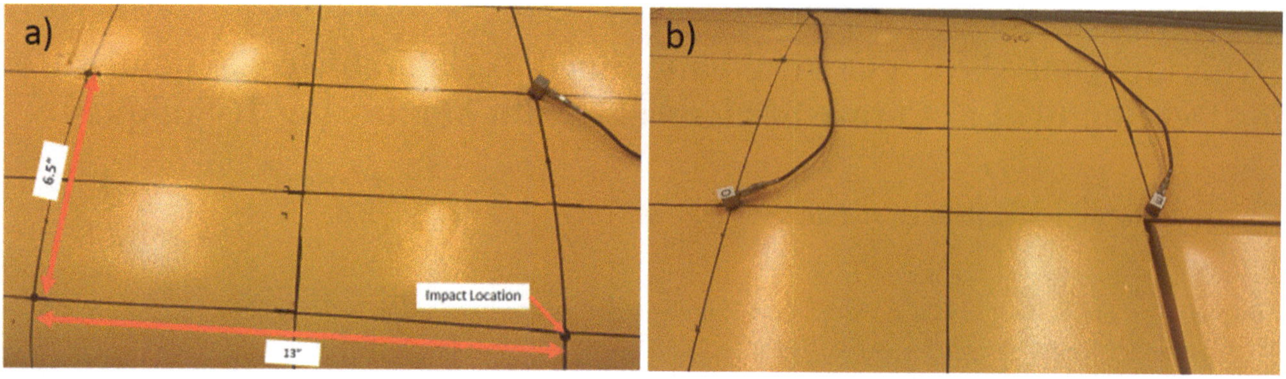

Fig. 10.1 (**a**) Image if grid size accelerometer attached to airfoil for fixed response testing (**b**) accelerometers for fixed excitation tests

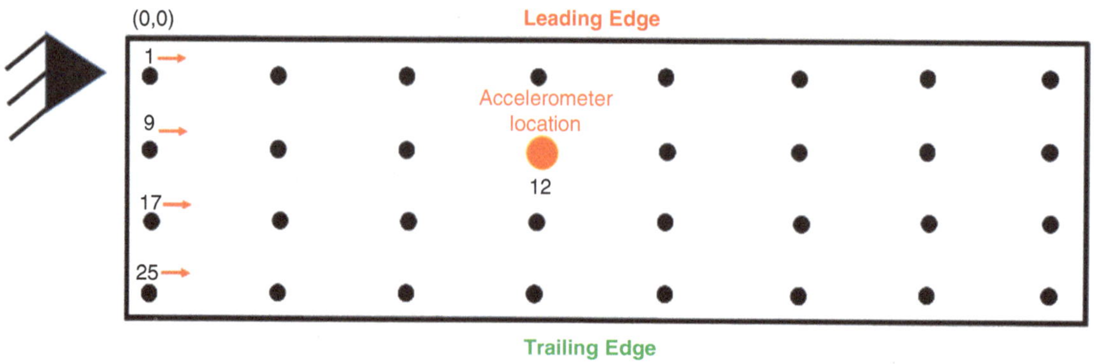

Fig. 10.2 2D representation of the UAV air foil for the fixed excitation modal tests

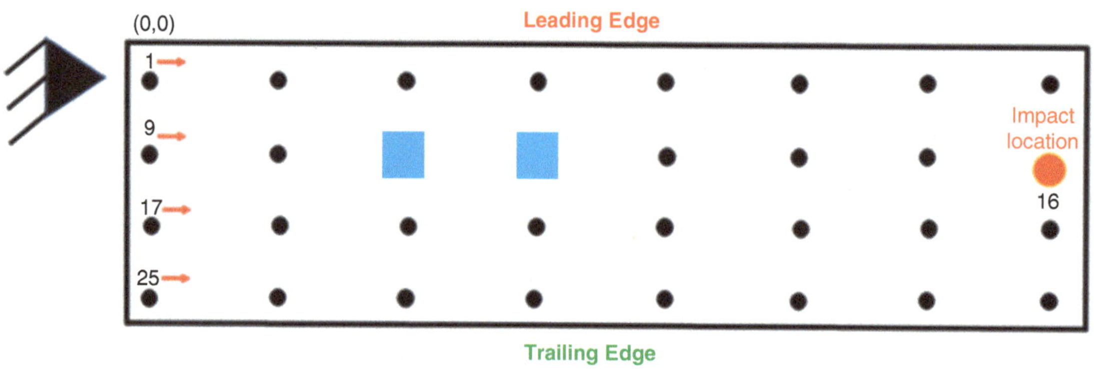

Fig. 10.3 2D representation of the UAV air foil for the fixed response tests

the modal frequency at the first peak of the frequency response curve is 61.16 Hz and it appears to be the bending mode (mode 1) response.

In terms of the fixed excitation tests, the torsional modal frequency at the first peak of the frequency response curve is 122.61 Hz, and it appears to be a mode 2 response. After investigating other responses in the fixed excitation tests, it is concluded that the there is evidence of a mixed modal response (mode 3) which could be contributed to the curvature of the wing. Future testing will be performed with a modal shaker that will supply constant excitation. This will allow for more consistent excitations throughout the wing and could possible give us more consistent mode shapes. Furthermore, Finite element analysis will be considered and can be used to compare to these physical testing parameters presented, as well as give insight on possible damage "hot spots" are throughout the wing under certain modes as well as mixed modes.

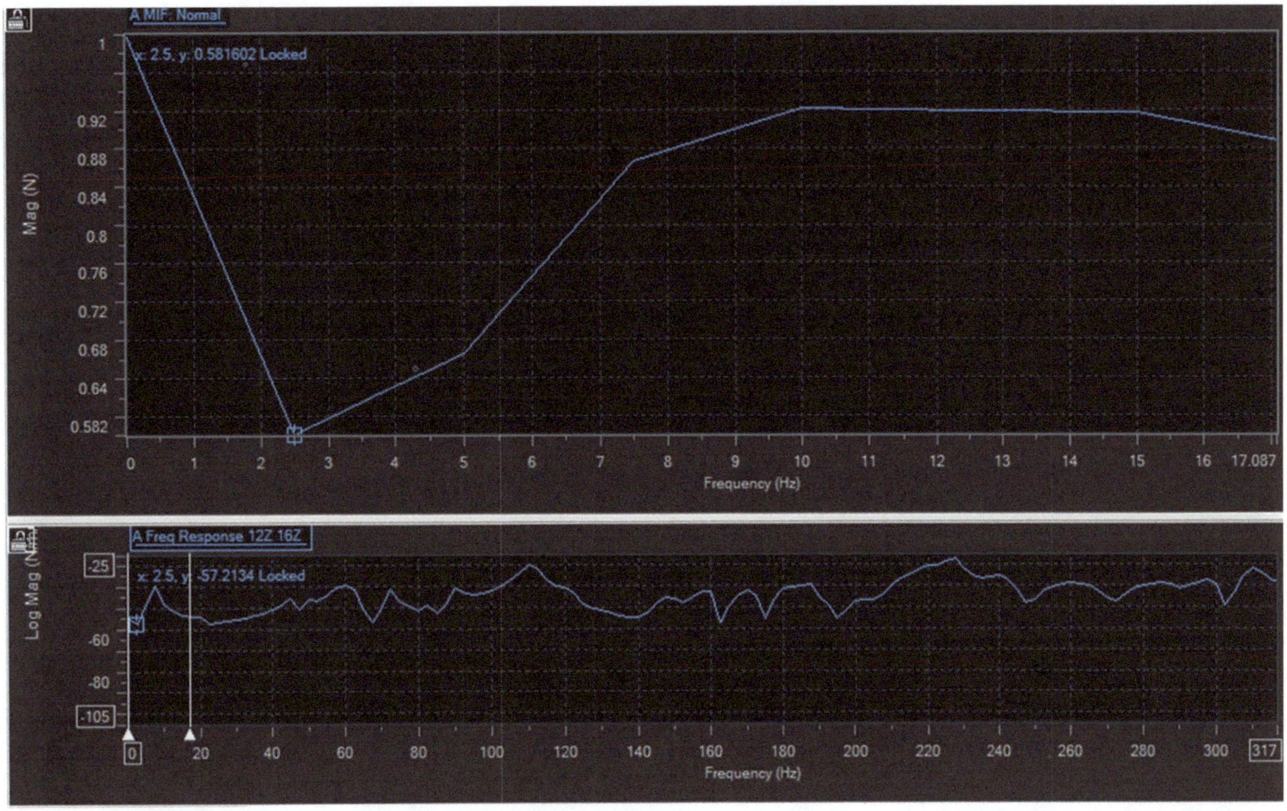

Fig. 10.4 Frequency response of a fixed response test at impact point 12

a b

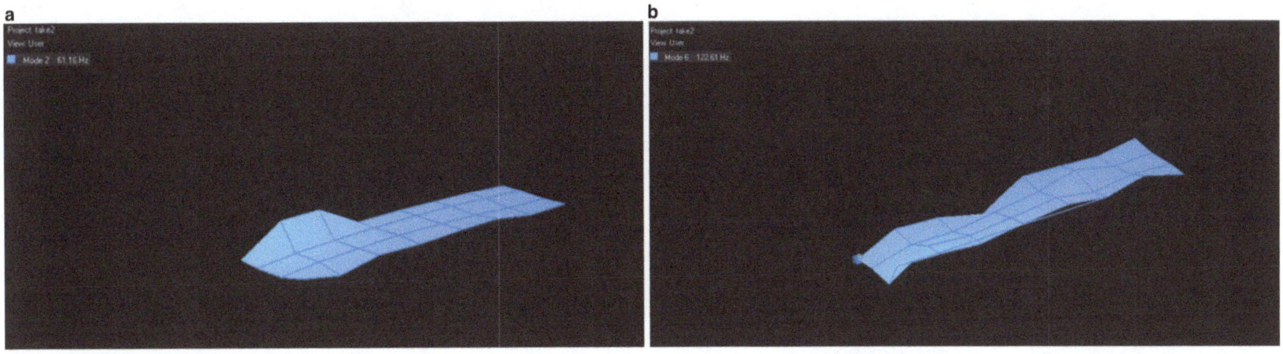

Fig. 10.5 (a) Modal image at the first peak of the modal response with a fixed response tests (b) Modal image of at the first peak of the frequency response of the fixed excitation modal tests

References

1. Lemler, K., Semke, W.: Application of Modal Testing and Analysis Techniques on a sUAV, School of Engineering and Mines, University of North Dakota, Grand Forks, ND 58202
2. Lanier, P.: Stereovision Correction Using Modal Analysis. Virginia Polytechnic Institute and State University, Blacksburg (2010)
3. Simsiriwong, J., Sullivan, W.R.: Vibration Testing of a Full – Scale Carbon Composite Ultralight Unmanned Aerial Vehicle, In: 51st AIAA/ASME/ASCE/AHS/ASC Structures, Structural Dynamics, and Materials Conference, Orlando, FL, 12–15 April 2010
4. Mastroddi, F., Giuliano, C., Andrea, C.: Aeroelastic Identification of a Flying UAVs by Output Only Data with Applications on Vibration Passive Control, In: 51st AIAA/ASME/ASCE/AHS/ASC Structures, Structural Dynamics, and Materials Conference, Orlando, FL, 12–15 April 2010
5. Subarano, K., Supekar, A., Lawrence, K.: Investigation of Morphable Wing Structures for Unmanned Aerial Vehicle Performance Augmentation, In: AIAA Unmanned...Unlimited Conference, Seattle, Washington, 6–9 April 2009

Chapter 11
Water Shock Experiments on Automotive Underbody Panels

Kevin A. Gardner, Jeremy D. Seidt, and Amos Gilat

Abstract Automotive underbody panels are susceptible to water shock loading that can be generated by driving over a puddle at high speeds. The thin panels can be permanently deformed or fail if the loading is large enough. Aluminum underbody panels are bolted into a custom test rig and subjected to water jet loading from a commercial pressure washer. The force on the panels is measured using four load cells and displacement data is captured using three-dimensional image correlation. The flow field is characterized using orthogonally mounted high speed cameras to determine the jet velocity. Additional experiments are performed using a custom water cannon that fires slugs of water contained in latex balloons at the underbody panels. This more accurately recreates the water splash event created when driving over a puddle. The velocity and shape of the water slugs are measured using orthogonally mounted cameras.

Keywords Fluid-Structure Interaction • Water Shock • Digital Image Correlation • Experimental Mechanics • Automotive

11.1 Introduction

Current releases of LS-DYNA [1] allow the user to study fluid-structure interactions with both the compressible and incompressible solvers. This provides a powerful design tool that can be used to help trim weight from automobiles. Automotive underbody panels are typically stamped from thin sheet metal or molded with polymers. These panels are intended to improve aerodynamics and provide protection for exposed components. Weight reduction is one of the simplest methods for improving fuel economy and can be accomplished by thinning the automotive panels. However, if underbody panels are made too thin they can become prone to deformation or failure from water shock loading. Driving over a puddle at high speeds causes a water shock that can hit the underbody panel directly or glance off of other components before hitting the panel. Experimentally validated results from fluid-structure interaction simulations can be used to predict and prevent the failure of underbody panels while optimizing their design to save weight and material.

Many researchers have studied fluid-structure interactions in different scenarios and loading configurations. Deformation of steel plates subjected to underwater impulsive loading was investigated using a custom test apparatus [2]. A large water chamber is incorporated into a gas gun to generate the loading. Test plates were bolted onto the end of the water chamber and deformations were tracked using shadow moiré techniques. The deformation of composite structural plates subjected to underwater blast loadings has also been investigated using a test setup that replicates common fluid-structure interactions [2–4].

11.2 Experimental

Automotive underbody panels are primarily subjected to water shock loadings when driving over a puddle at high speeds. The interaction of the tires with the puddle creates high velocity sprays and splashes that impact the underbody panel. In order to match conditions observed by production vehicles and eliminate the need for fabricating lab scale test panels, full size aluminum underbody panels were using during the testing. The aluminum panels were bolted into the test rig using existing mounting holes. The panel holder was constructed of solid steel box beam and was vertically supported by linear bearings.

K.A. Gardner (✉) • J.D. Seidt • A. Gilat
The Ohio State University, Department of Mechanical and Aerospace Engineering, Columbus, OH, USA
e-mail: kevin.gardner.796@gmail.com

P. Zavattieri et al. (eds.), *Mechanics of Biological Systems, Materials and other topics in Experimental and Applied Mechanics, Volume 4*, Conference Proceedings of the Society for Experimental Mechanics Series, DOI 10.1007/978-3-319-63552-1_11

Fig. 11.1 Test rig

Fig. 11.2 Water cannon

Load cells were placed between the panel holder and the main support stand to record the forces acting on the panels. The entire apparatus was placed within a catch basin to contain the water and an acrylic enclosure was used to contain the overspray. The clear acrylic allowed the use of digital image correlation to measure the deformation in the panels. A section of optically clear Lexan was used for the Photron SA1.1 cameras used for the image correlation. Additional high speed cameras were also mounted orthogonally to characterize the water jet and splashes used to load the panels. The test rig is shown in Fig. 11.1.

During testing the underbody panels were loading with a water jet and splashes. The water jet was created using a commercial pressure washer. The pressure washer was rated at 10 HP, 4000 psi, and 3.5 gpm. A water reservoir was constructed using a 100 L water barrel and a 1 HP transfer pump, rated at 25 gpm, was used to provide a pressure head to the pressure washer. The spray wand of the pressure washer was clamped in place and supported by the test rig to ensure that it did not move when the trigger was depressed. Clamping the spray wand in place also allowed the standoff distance and loading location to be consistent for all panels tested.

The splash loading was generated using a custom water cannon that fired slugs of water contained within latex balloons. The water cannon, shown in Fig. 11.2, was constructed from two 24 inch sections of three inch NPT pipe. One side acted as the pressure chamber and the other side was the barrel. A three inch butterfly valve was used to control the firing of the water cannon. The support stand for the water cannon positioned the center of the barrel at the same height as the pressure washer nozzle so the panels could be loaded in the same location as the water jet.

Fig. 11.3 Sabot and water balloons

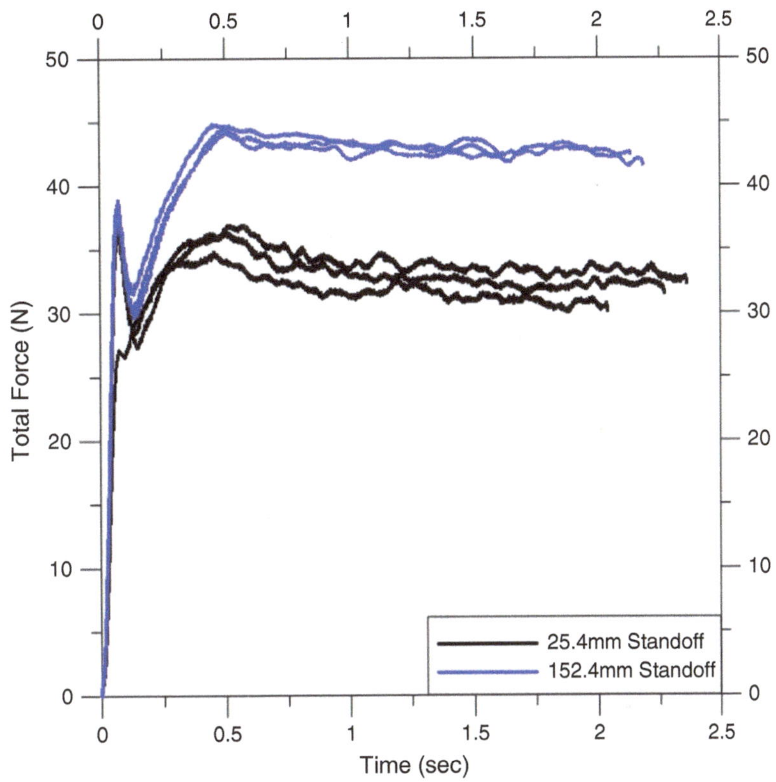

Fig. 11.4 Typical force history

The filled water balloons were placed within a sabot made from high density polyethylene as shown in Fig. 11.3. Each balloon was filled with 120 mL of tap water. The water balloons fit snugly into the sabot when filled. The balloons were lubricated before being placed in the sabot to create a hydraulic seal with the sabot. This caused the balloons to rupture when the sabot was stopped by the end of the barrel, generating the splash loading.

Four underbody panels were tested using the water jet loading. Two standoff distances of 25.4 mm and 152.4 mm were chosen and a 0° spray nozzle was used for all tests. Each test configuration was repeated three times. The larger standoff distance was tested first for each panel since minor pitting occurred at the loading point for the shorter standoff distance. Figure 11.4 shows typical force histories recorded for one of the test panels. The total force acting on the panel in the axial direction was determined by summing the measurements from the four load cells between the panel holder and support stand. The total force was lower with the shorter standoff distance due to the geometry of the panel. The panels had reinforcing ribs and at the shorter standoff distance a larger percentage of the flow's momentum glanced off in various directions because the cross-sectional area of the stream was less than for the larger standoff distance. This effect was corroborated by tests performed on a flat plate of aluminum. Figure 11.5 shows the force vs. displacement data for all tests performed with a 25.4 mm standoff distance.

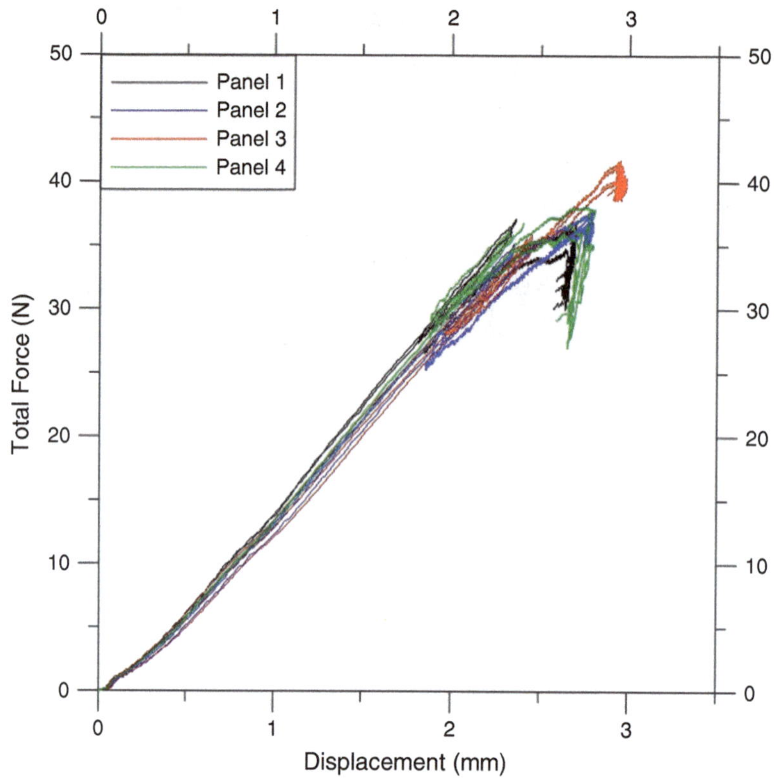

Fig. 11.5 Force vs. displacement

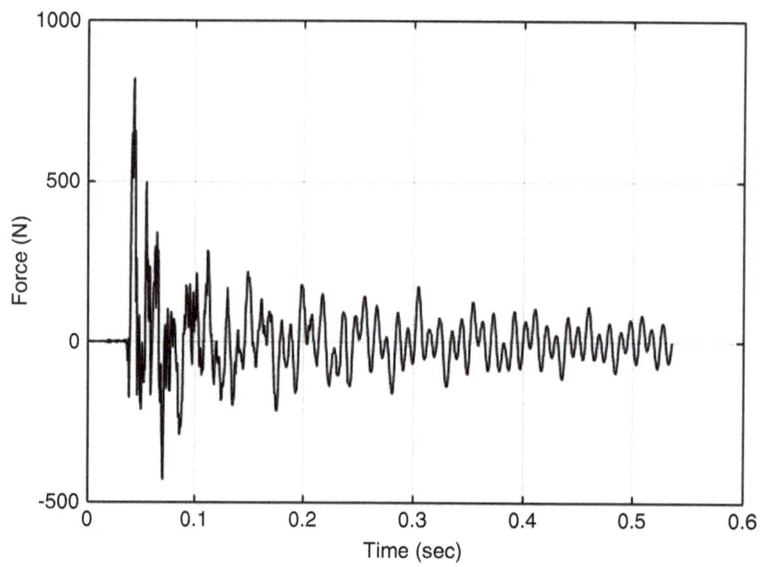

Fig. 11.6 Typical force history

Each of the panels used for the water jet experiments was tested using the water cannon. Each panel was tested using a reservoir pressure of 207 kPa which resulted in an average velocity of 33 m/s for the water splash. The velocity of the splash was determined using orthogonally mounted high speed cameras and a reference scale placed in the splash path. The water splash resulted in significantly higher forces than the water jet and each of the test panels were permanently deformed. Figure 11.6 shows a typical force history for the water cannon testing. Figure 11.7 shows a time lapse history of a typical splash generated by the water cannon.

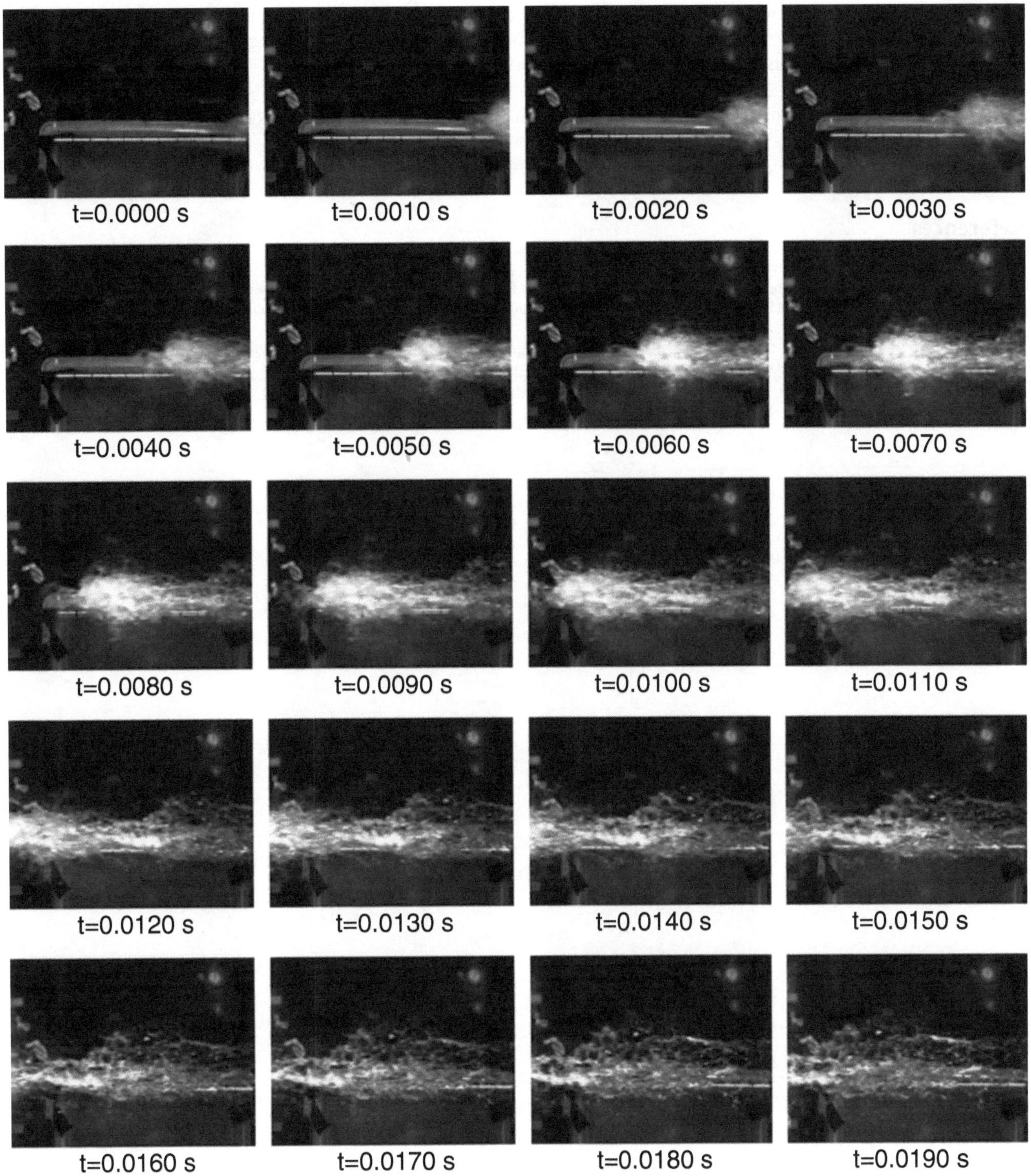

Fig. 11.7 Typical time history of splash

11.3 Summary and Conclusions

An experimental test procedure was presented to test automotive underbody panel under water shock loading. A test rig was designed to allow testing on full size panels. The panels were tested with water jet loading from a commercial pressure washer and splash loading from a custom water cannon. Test data indicates variations between the panels due to variability in the forming operation. The water jet testing did not exhibit any permanent deformation of the panels, however the higher loading forces from the water splash tests did result in permanent deformation. These experimental data will be used to validate fluid-structure interaction simulations in LS-DYNA.

Acknowledgements The authors would like to thank Duane Detwiler, Andy Fields, Allen Sheldon, Thomas Ramsay, and Yong-Bae Cho of Honda R&D Americas, Inc. and Ed Helwig and Emily Nutwell of LSTC for their support of this project.

References

1. Pin, F.D.: Advances on the Incompressible CFD Solver in LS-DYNA. In: 11th International LS-DYNA Users Conference
2. Espinosa, H.D., Lee, S., Moldovan, N.: A novel fluid structure interaction experiment to investigate deformation of structural elements subjected to impulsive loading. Exp. Mech. **46**, 805–824 (2006)
3. Wei, X., Tran, P., Vaucorbeil, A., Ramaswamy, R.B., Latourte, F., Espinosa, H.D.: Three-dimensional numerical Modeling of composite panels subjected to underwater blast. J. Mech. Phys. Solids. **61**, 1319–1336 (2013)
4. Wei, X., Vaucorbeil, A., Tran, P., Espinosa, H.D.: A new rate-dependent unidirectional composite model - application to panels subjected to underwater blast. J. Mech. Phys. Solids. **61**, 1305–1318 (2013)

Chapter 12
Experimentally Validated Finite Element Simulation of Aluminum Extrusion of a Micro-Multiport Condenser

Tushar Bakhtiani, Hazim El-Mounayri, and Jing Zhang

Abstract The research aims at developing and experimentally validating a finite element (FE) model that can be used to optimize the geometry of a porthole die for aluminum extrusion to maximize tooling life while achieving the required product quality. A commercial FEA code that was validated in another study, namely DEFORM 3D®, is used to create a thermo-mechanical model that closely represents the highly non-linear large deformation nature in the extrusion process. Both Lagrangian formulation and Eulerian formulation are explored. Using the Updated Lagrange FE analysis, the tooling components of the modular die are assumed to be rigid in order to achieve the right friction coefficients to validate the experimental data collected during trail runs of the modular die. A steady state FE analysis is run to predict the maximum weld pressure between the mandrel port webs. The simulation is used to study the effects of port opening, port lead angle, and angle intercept distance, on effective stress, material velocity, and welding pressure. Different parameters are predicted, including temperature distribution, wear, stress and stress concentration, extrusion load, exit velocity, exit temperature, material flow through the die port, etc. The insight gained from these simulations can be used to optimize the geometry of the die mandrel.

Keywords Extrusion Die • FE Simulation • DOE • Optimization • DEFORM 3D®

12.1 Introduction

Aluminum extrusion is a complex thermo-mechanical process during which the flow of the deforming aluminum billet is primarily dictated by frictional forces at the billet container and tooling interface. Bulk deformation of the aluminum billet occurs in the container as compared to the die cavity therefore, frictional forces at the billet and container interface are of utmost importance. Due to high temperatures and pressures occurring during the extrusion process, it is difficult to obtain experimental data of friction. As a result, friction mechanism during aluminum extrusion has not been thoroughly understood and mathematically expressed, therefore it has been acknowledged as the most uncertain parameter in the finite element simulation for aluminum extrusion [1]. The main research aim of this work is to develop and experimentally validate a finite element (FE) model that can be used to optimize the geometry of a porthole die for aluminum extrusion to maximize tooling life while achieving the required product quality.

12.1.1 Finite Element Model Formulation

In order to accurately predict the material flow during a forming process, the formulation must consider large plastic deformation, incompressibility, interface contact between tooling and deforming metal and temperature. The FEM formulation in DEFORM 3D® is based on minimum work rate principle [2, 3], which is determined using the variational principle equation shown by mixed formulation Eq. (12.1), employed for the 3D tetrahedral elements to obtain the variable solution for velocity and pressure.

T. Bakhtiani • H. El-Mounayri (✉) • J. Zhang
Department of Mechanical Engineering, Purdue School of Engineering and Technology, Indiana University – Purdue University Indianapolis, 723 West Michigan Street, SL 260, Indianapolis, IN, 46202, USA
e-mail: helmouna@iupui.edu

P. Zavattieri et al. (eds.), *Mechanics of Biological Systems, Materials and other topics in Experimental and Applied Mechanics, Volume 4*, Conference Proceedings of the Society for Experimental Mechanics Series, DOI 10.1007/978-3-319-63552-1_12

$$\delta\varnothing(v,p) = \int_V \bar{\sigma}\delta\dot{\bar{\varepsilon}}dV + \int_V p\delta\dot{\varepsilon}_V dV + \int_V \dot{\varepsilon}_V \delta p - \int_S F_i \delta v_i dS = 0 \tag{12.1}$$

where $\int_V \bar{\sigma}\delta\dot{\bar{\varepsilon}}dV$ is the variational form of the plastic work done $\int_V p\delta\dot{\varepsilon}_V dV + \int_V \dot{\varepsilon}_V \delta p$ is the dialation work rate integrated by parts, which together form the constant volume constraint and $\int_S F_i \delta v_i dS$ is the surface traction work. $\dot{\varepsilon}_V$ is the volumetric strain rate, $\bar{\sigma}$ is effective stress, $\dot{\bar{\varepsilon}}$ is the effective strain rate, p is the pressure, V is the volume of the deforming material and S is the surface of the deforming material. This variational equation is converted into simultaneous non-linear algebraic equations utilizing FEM discretization procedure which is solved numerical either using direct iteration method or Newton-Raphson.

Temperature is an important characteristic of the extrusion process that influences the material flow stress. Therefore, the temperature distribution occurring between the deforming material and die tooling can be obtained by rewriting the thermal energy balance equation using weighted residual method shown by Eq. (12.2):

$$\int_V kT_j \delta T_j dV + \int_V \rho c\dot{T}\delta\dot{T}dV - \int_V \alpha\bar{\sigma}\dot{\bar{\varepsilon}}\delta T dV - \int_S q_n \delta T dS = 0 \tag{12.2}$$

where $\int_V kT_j \delta T_j dV$ is the conductive term, $\int_V \rho c\dot{T}\delta\dot{T}dV$ is the heat capacity term, $\int_V \alpha\bar{\sigma}\dot{\bar{\varepsilon}}\delta T dV$ is the heat lost due to plastic deformation, $\int_S q_n \delta T dS$ is the heat flux occurring between the deforming material and die tooling components, k is the thermal conductivity, T is the temperature, ρ is the density of the deforming material, c is the specific heat, α is the percentage of deformation energy that converts into heat and q_n is the heat flux normal to the boundary.

12.1.2 *Friction Model*

Much experimental research and theoretical modelling has been conducted to understand the friction phenomenon occurring during the extrusion process. However, assigning accurate frictional coefficients and models is one of the key issues for the finite element analysis of the extrusion simulations [4–7]. Shear and Coulomb friction, are the two of the oldest and widely used classical frictional models in most commercial FE codes, due to their theoretical and numerical simplicity.

The coulombs frictional model can be described as a function of normal pressure between two contacting bodies:

$$f = \mu p \tag{12.3}$$

where f is frictional stress, μ is the friction coefficient and p is the interface pressure between two bodies. The shear friction model can be described as a function of the yield stress of the deforming body:

$$f = mk \tag{12.4}$$

where f is frictional stress, m is the friction coefficient and p is the shear yield stress of the deforming body.

Both friction models are equally popular and have been utilized by many researchers. Sofuoglu et al. [8], and Arentoft et al. [9], among many others have successfully utilized the coulomb friction model. Xianghong et al. [10] and Khansai et al. [11], among many others have successfully utilized the shear friction model. Both of these frictional models are numerically implemented and readily available to the user in deform. However, selection of the friction coefficient is based on an educated guess [4] as it is not easy to determine these values through experimentation due to the dynamics of the aluminum extrusion process.

Therefore, for the comparative study the shear and coulomb friction models are both evaluated and can be seen in Table 12.1.

Table 12.1 Comparative study of friction coefficients

	Friction coefficient	
Case	Billet Container interface	Billet Tooling interface
1	Coulomb 0.9	Coulomb 0.4
2	Shear 0.9	Shear 0.4

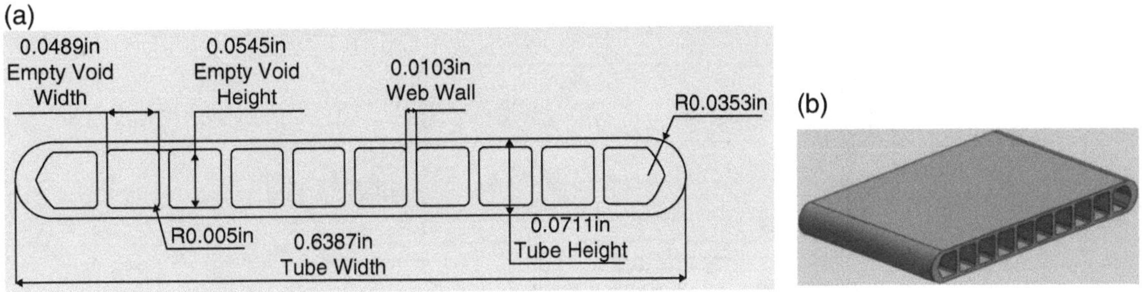

Fig. 12.1 Micro-Multiport tube (MMP) (**a**) tubing dimension and (**b**) 3D profile

The coulomb friction parameters for case 1 were provided by a research scientist from a local extrusion company who had based these parameters from his experience in the industry and empirical knowledge of conducting extrusion simulations. The shear friction parameters in case 2, was referenced from the works of Jian et al. [12].

12.1.3 Validation Criteria

Contemporary heat exchange applications such as air conditioning systems, evaporators, radiators and refrigerators use parallel flow tubes that are fabricated with aluminum extruded Micro-Multiport (MMP) tubes. This tubing, which is also referred to as condenser tubes, is typically made out of 1000 or 3000 series aluminum alloys due to their characteristic properties of high performance, low weight, low corrosion, and high recycling value. In general, the MMP tube is a flat body with a row of side-by-side passageways, which are separated by upright web walls as shown in Fig. 12.1. A local extrusion company that has previously used the modular die for the extrusion of MMP tubes has provided the initial input parameters of the extrusion process as well as experimental data of the exit temperature and extrusion load. Since, friction is the only unknown parameter a comparative study with different friction parameters is conducted to validate the accuracy of the simulated results with the experimental data provided.

Figure 12.2a shows the extrusion force required by the hydraulic extrusion press to push multiple billets through the porthole die tooling assembly that consist of eight die cavities. The extrusion process is a continuous process; therefore, the peaks shown in the figure for maximum extrusion force are considered to be the breakthrough forces that marks the onset of steady state extrusion after which the maximum extrusion forces reduces gradually as the length of the billet decreases. In this study the extrusion of the first billet is simulated therefore, we compare the simulated results with the experimental data for the first few billets which shows the first peak of maximum extrusion at 2400 tons. Similarly Fig. 12.2b shows the temperature of the MMP tube extruded from the porthole die which is measured after the tube exits the die through an infrared thermometer. The maximum temperature in the experimental data of the first few billet is close to 1150 °F which is also considered as the validation criteria for the simulated results.

12.1.4 Description of the Validation Model and Mandrel Model

12.1.4.1 Validation Model: Micro-Multiport Tubing (MMT)

In the MMT validation model, the billet was modelled as a plastically deforming 3003 series aluminum alloy whereas, the porthole die tool stack, container and ram were defined as rigid H13 tool steel material. The extrusion ram speed was defined at 6 mm/s; whereas, initial temperature of the billet and tooling was defined at 900 °F, respectively. The heat transfer coefficient are readily available in DEFORM 3D® for forming and free resting objects. Therefore, the heat transfer coefficient between the forming billet and die tooling components is 0.002 $\frac{Btu}{\sec in^2 {}^{\circ}F}$ and the heat transfer coefficient between resting objects which is the tooling components and die is 0.0003 $\frac{Btu}{\sec in^2 {}^{\circ}F}$. In addition, symmetric planes were assumed to be fixed with no material moving across due to the use of one-sixteenth of the model. The simulation was setup to stop after the ram has been pushed down 0.45 inch, and the simulation information was setup to save every ten steps. The length of the simulation step was defined in terms of ram displacement, and each step is defined as a constant value of 0.01 in/step. The

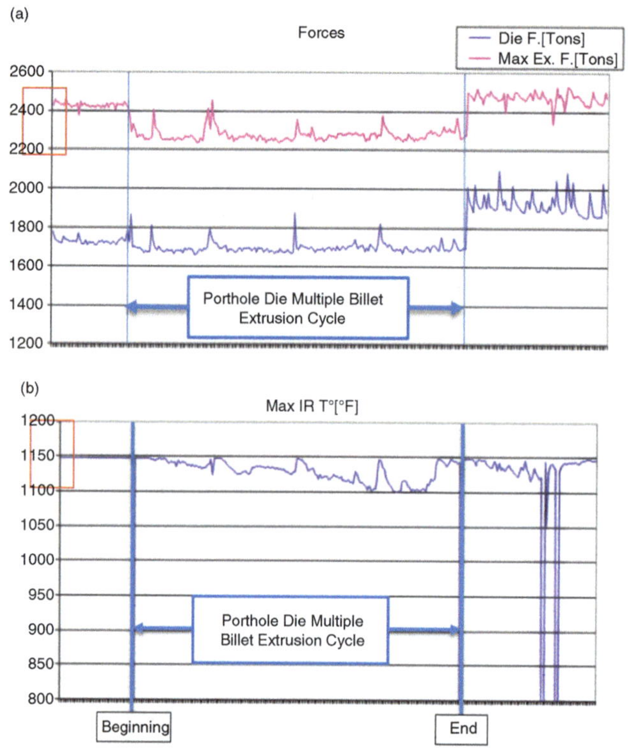

Fig. 12.2 (**a**) Production extrusion force on ram for multiple billet extrusion cycles; (**b**) Production extruded MMP tube temperature

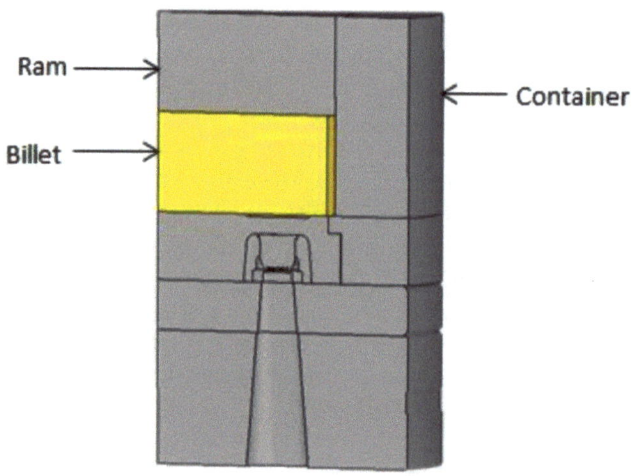

Fig. 12.3 Modular die FEA Model

diameter of the billet used in the simulation is 8 inches and the inner diameter of the container was set at 8.375 inches. This was done to mimic the upsetting stage of the extrusion process.

12.1.4.2 Aluminum Extrusion Model: Mandrel

One sixteenth of the modular die geometry was utilized for this comparative study validation. In the actual extrusion process the length of a single billet is 40 inches which would need a lot more number of elements to model. Therefore, to save time and computational costs (reducing mesh and total number of elements) only a 4 inch long aluminium billet was simulated as shown in Fig. 12.3.

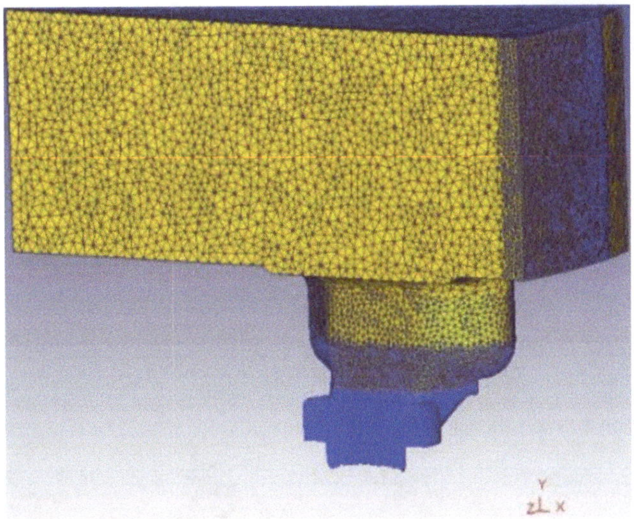

Fig. 12.4 Deformed billet with several mesh window

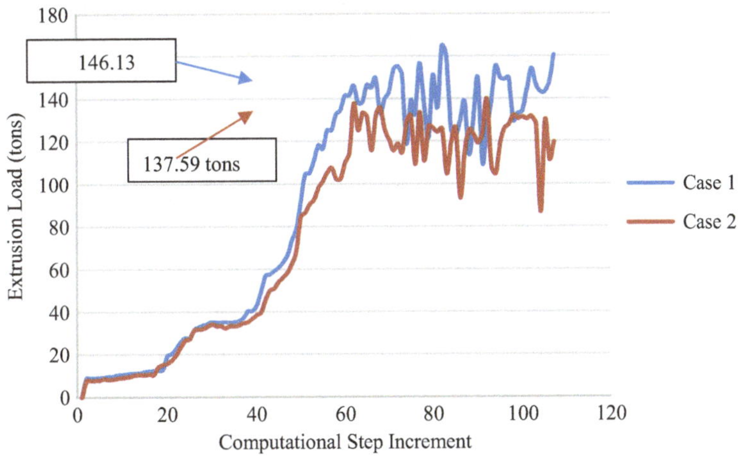

Fig. 12.5 DEFORM 3D® simulation extrusion load curve

As the ram forces the billet through the porthole cavity the billet will undergo severe plastic deformation due to the reduction in cross sectional area. Therefore; to control the element number and ensure calculation accuracy; several local mesh windows with a higher element density are applied on the billet in accordance with the extent of local deformation shown in Fig. 12.4.

A windows operating with 32GB of RAM and four 3.4Ghz Intel Core i7 processors was utilized to conduct the numerical simulation of the extrusion process in DEFORM 3D®. The total computational time for each simulation is approximately 16 hours mainly due to re-meshing of the billet due to large scale deformation.

12.1.5 Validation Model Results

12.1.5.1 Effect of Friction on Extrusion Load

Figure 12.5 compares the extrusion load stroke of coulomb and shear friction models. The extrusion curves observed in Fig. 12.8 are not smooth and show fluctuations at each step after the breakthrough pressure, this is due to the use of UL formulation in DEFORM 3D®, that causes multiple re-meshes of the deforming billet once the elements reach the small dimension of the MMP tube web walls in mandrel port opening. However, from the simulated results it is observed that the breakthrough force occurs at the same computational step increment and is 146.13 tons for trial 1 and 137.59 tons for trial

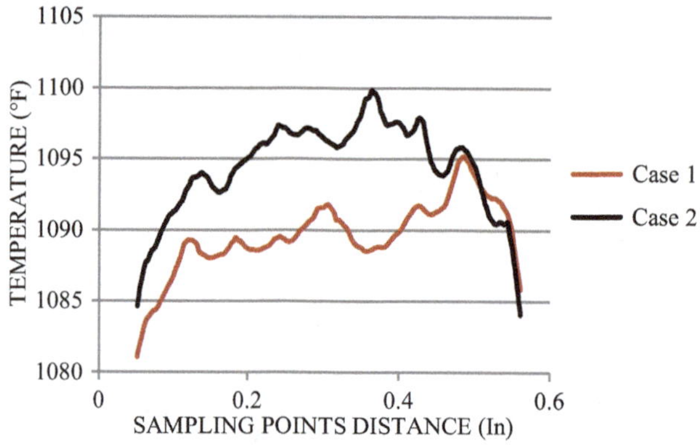

Fig. 12.6 Exit temperature measured on the outer surface of MMP tube

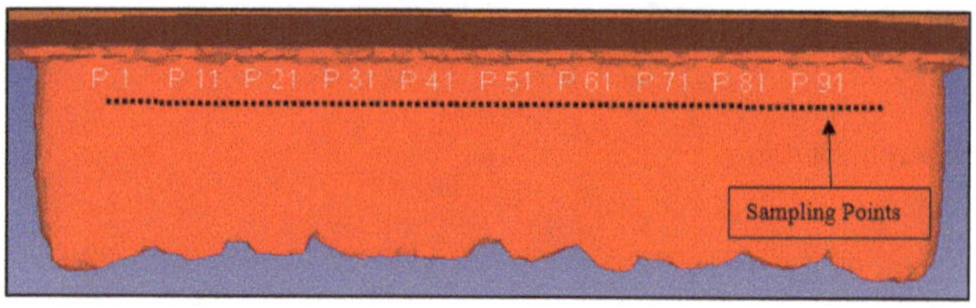

Fig. 12.7 Exit temperature sampling points on surface on MMP Tube

Table 12.2 Friction model validation results

Case	Friction parameter		Max extrusion (Tons)	% Load difference	Max temperature (Degree °F)	% Temperature difference
	Billet container interface	Billet tooling interface				
1	Coulomb 0.9	Coulomb 0.4	2338	2.667%	1090	5.333%
2	Shear 0.9	Shear 0.4	2201	8.602%	1095	4.888%

2. The simulated results of extrusion load is for one sixteenth of the actual porthole die model, which is approximately 2338 tons for trial 1 and 2201 tons for trial 2 of the entire porthole die model due to symmetry.

The results obtained from this simulation is only for the first billet cycle; therefore, when comparing these results to the multiple billet cycle experimental data, we compare the simulated breakthrough loads with the average value of the experimental extrusion load observed in Fig. 12.4, which is 2325 Tons. Both the simulated results are in good agreement with the experimental data provided. However, coulomb friction present in trial 1 has a breakthrough pressure close to the average value of the experimental extrusion load.

Effect of Friction on Exit Temperature

Figure 12.6 shows the temperature distribution obtained from each simulation trial. The data represented in Fig. 12.6 was calculated using sampling points along the outer wall of the condenser tube as shown in Fig. 12.7.

From the graphs of the exit temperature measured on the surface of the MMP tube. It is observed that the difference in exit temperature is very minute, as the exit temperature in trial 1 is close to 1090 °F and trial 2 is 1095 °F. Friction greatly influence the flow of aluminum, extrusion load and temperature in the extrusion process where bulk deformation occurs; i.e. where contact is more intimate and pressure are significantly higher, such as the container and porthole [6, 13]. However, the same is not true in the bearing region as this region is characterized by maximum extrusion velocity and minimum pressure due to the drastic reduction in cross sectional area. As a result the two frictional models in consideration lose their influence in the bearing region due to its small size. The results from both the simulations are below the exit temperature values obtained in the experimental data, due to the assumed value of heat transfer coefficient.

A summary of the comparative study results can be seen in the Table 12.2. Both the simulations results are in good agreement with the experimental data.

12.2 Design of Experiments of the Mandrel

The mandrel is one of the critical components in the modular die design. The function of a mandrel is to help create the thin web walls on the insides of the MMP tube that are separated by fine voids. The Port Opening, Port Lead Angle and Port Angle Intercept Distance are the key characteristics of the Mandrel CAD geometry that allow the formation of the thin web walls of the MMP tube (Fig. 12.8).

The port opening feature was designed to allow sufficient amount of aluminum to flow in between the teeth's to form a good weld before the actual dimension of the web walls are defined in the bearing (Fig. 12.9).

The port lead angle and port angle intercept distance was designed with the intuition that the aluminum would flow along the angled walls and start welding at the specified port angle intercept distance. These three parameters of the mandrel help in forming the port geometry that allows to increase the amount of aluminum flowing into the porting geometry. Therefore, using a full factorial DOE approach, three factors (Port opening, port lead angle and intercept distance) each at three levels were statistically analyzed to understand the effects of the mandrel CAD geometries on aluminum flow into the porting geometry in order to maximize weld pressure.

Due to the linear design of the mandrel a simplified cross sectional model was utilized to conduct the 27 mandrel DOE's, as shown in Table 12.3. The three features under investigation are consistent between each mandrel tooth. Therefore; this model was created focusing only on one port opening to help reduce computation time. The cross sectional model consisted of 200,000 elements in the cavity filled work piece; which consisted of multiple mesh windows of higher mesh density to gain good accuracy in results of the critical features under investigation. All the tooling components were considered to be

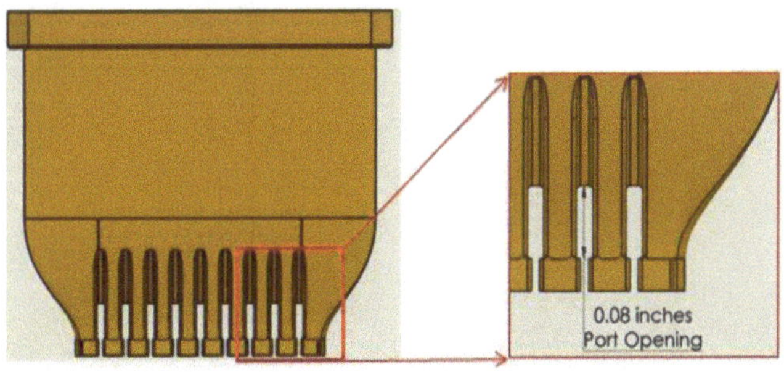

Fig. 12.8 Port opening in Mandrel

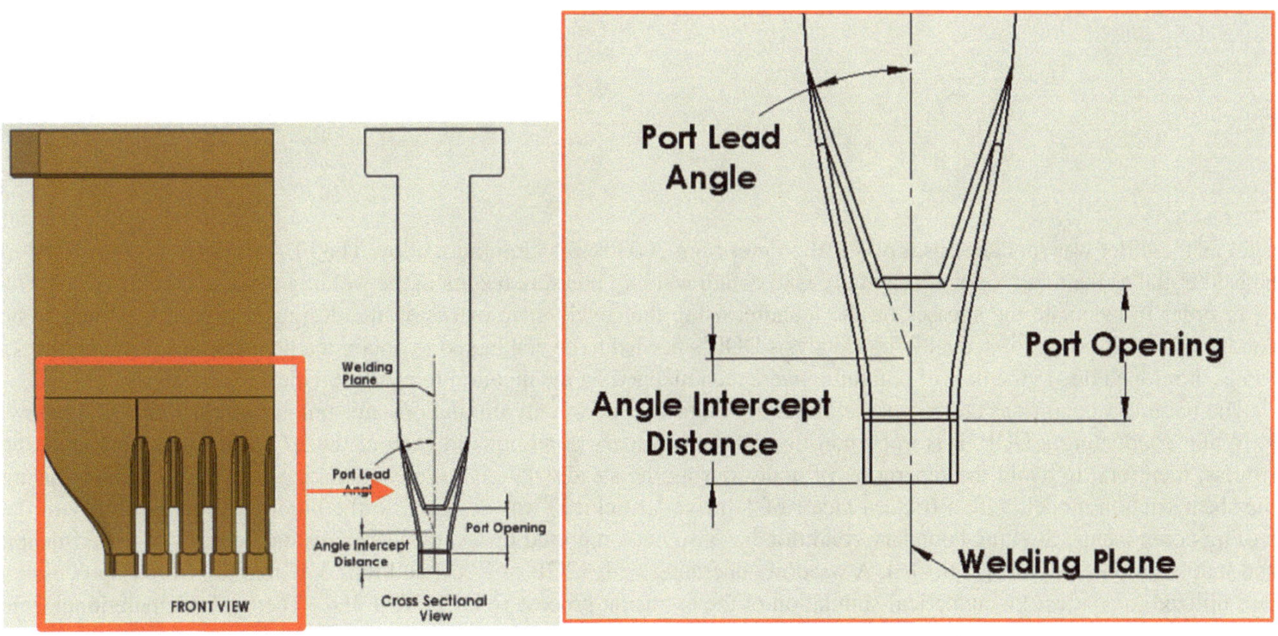

Fig. 12.9 Cross sectional view of the three parameters that influence aluminum flow into the porting geometry distance

Table 12.3 Mandrel full factorial design of experiments

Trial	Port opening	Port lead angle	Intercept distance
1	0.06	15	0.04
2			0.07
3			0.1
4		20	0.04
5			0.07
6			0.1
7		25	0.04
8			0.07
9			0.1
10	0.07	15	0.04
11			0.07
12			0.1
13		20	0.04
14			0.07
15			0.1
16		25	0.04
17			0.07
18			0.1
19	0.08	15	0.04
20			0.07
21			0.1
22		20	0.04
23			0.07
24			0.1
25		25	0.04
26			0.07
27			0.1

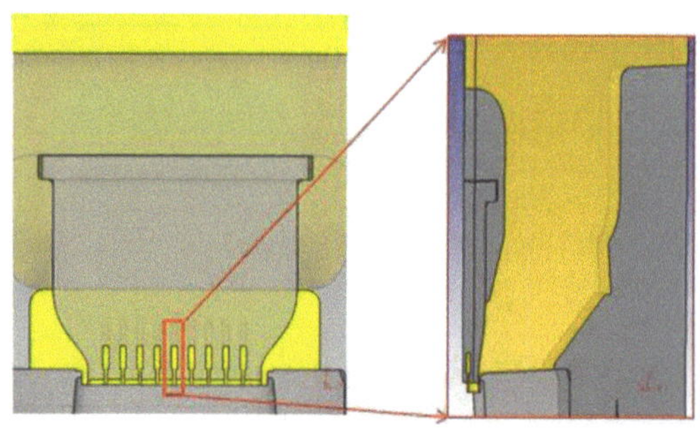

Fig. 12.10 Cross sectional Mandrel DOE model

rigid as the billet was modelled as a plastically deforming 3003 series aluminum alloy. The FEA model was considered to be isothermal as the main focus of this study is to obtain welding pressure results in the welding plane (Fig. 12.10).

In order to calculate the stresses on the mandrel using the steady state extrusion the design of experiments had to be conducted in two parts: First, the 27 flow analysis DOE's needed to be conducted to obtain the flow stresses. Second, the 27 forces developed due to the flow of aluminum were then mapped on the mandrel to conduct a die stress analysis.

The boundary conditions utilized in the Mandrel design of experiments simulations are summarized in Table 12.4 below:

While conducting a DOE, it is important that all the boundary conditions in each of the 27 DOE simulations must be similar; therefore, to avoid the separation of nodes during the steady state iterations a sticking boundary condition using the shear friction model, with a friction factor of 1 (m = 1, sticking) was assigned to the billet mesh in contact with the tooling components. Sticking boundary condition has also been reported in several literatures on the basis of experimental and industrial observations [4, 14–16]. A windows operating with 32GB of RAM and four 3.4Ghz Intel Core i7 processors was utilized to conduct the numerical simulation of the extrusion process in DEFORM 3D®. The total computational time for each simulation is 50 minutes.

Table 12.4 DOE process parameters

FE analysis conditions	
Extrusion condition	Value
Material of billet	AA 3003
Material of tools	H-13 tool steel
Billet temperature	900 C
Tooling temperature	900 C
Ram velocity	1.034 in/s
Friction conditions	
Tooling & billet interface	Shear 1
Mandrel & billet interface	Coulomb 0.4
Die plate & billet interface	Coulomb 0.4
Ram & billet interface	Shear 1

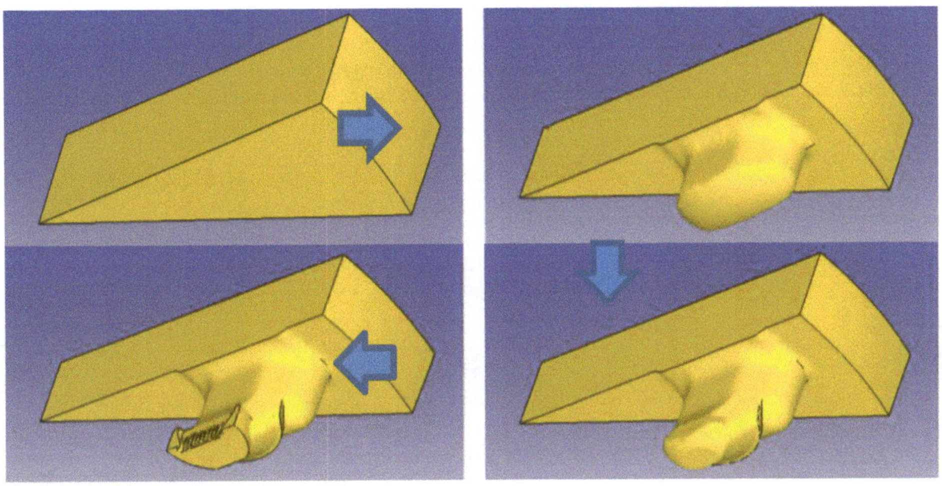

Fig. 12.11 Modular Die three dimensional forming process

12.3 Result and Discussion

12.3.1 Extrusion Force and Material Flow

The material flow of aluminum through the modular die can be categorized into four stages: dividing, filling, welding and forming stage (Figs. 12.11 and 12.12).

As the aluminum flows through the modular die, the cross sectional area of the die cavity reduces gradually which leads to an increase in the amount of load the ram has to exert on the deforming billet. This increase in load marks the beginning of each stage. First, increase in load is marked by the splitting of the billet into the porthole which gradual increase as the aluminum fills up the porthole. Second, filling stage, the load of the ram increase even further due to reduction in cross sectional area as the aluminum is forced to flow towards the mandrel. Third, Welding Stage, the divided aluminum flow will weld back together in the port opening of the mandrel to form the thin web walls of the MMP tube. Fourth, forming stage is marked by the peak value of extrusion load which forms the final shape of the MMP tube in the bearing area.

The welding of aluminum for the MMP tube initiates at the welding stage and is fully developed by the end of the forming stage where maximum force is exerted on the die by the aluminum flow. Therefore, to save computational time both, the mandrel and die cartridge design of experiments were simulated utilizing the steady state extrusion method in DEFORM 3D®, which accurately models the extrusion process during the forming stage of the MMP tube. The results of the DOE's were then analyzed using Minitab, a statistical software package that provides several analytical and graphical tools to help understand the results.

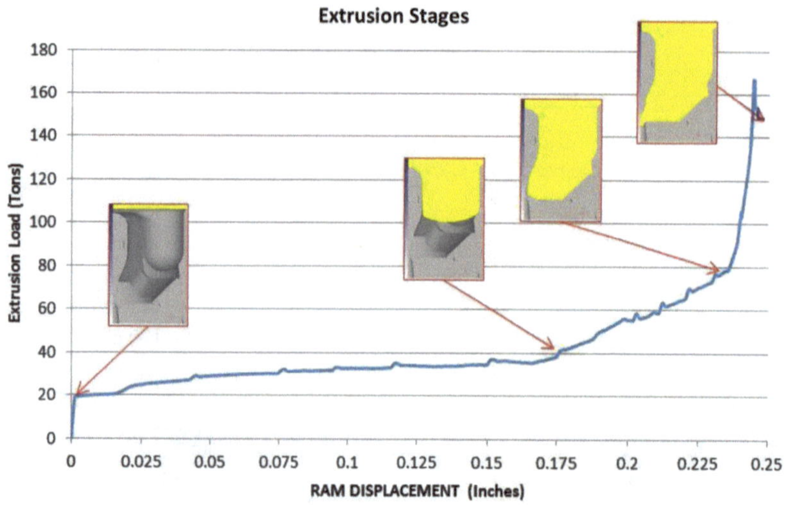

Fig. 12.12 Modular die extrusion load

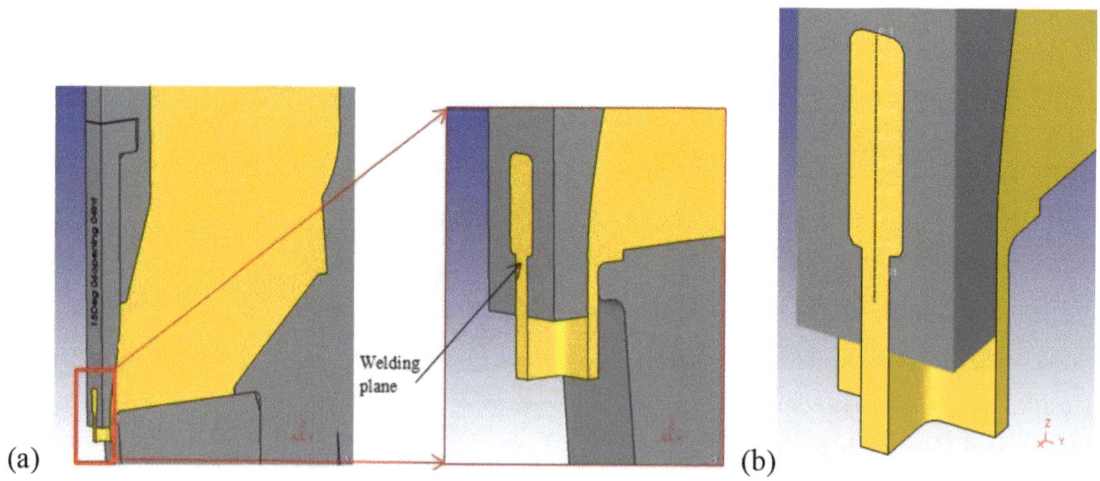

Fig. 12.13 Mandrel DOE (**a**) cross sectional model & (**b**) 100 sampling points on welding plane of port opening

12.3.2 Mandrel DOE Results

The mandrel design of experiments were simulated using the cross sectional model established earlier in the chapter. The main objective of this study is to investigate the effects of port opening; port lead angle and port angle intercept distance on weld pressure, material velocity and effective stress of aluminum on the welding plane. In order to measure the data on the welding plane, 100 equally distant sampling points were used form the entrance of the welding surface (Fig. 12.13).

12.3.2.1 Validation of Welding Pressure

In order to validate the results of the mandrel DOE, the distribution of welding pressure, and material velocity on the welding surface from one of the DOE simulations has been plotted and compared to the results Donati et al. [17] and LIU et al. [12] (Fig. 12.14).

The effective stress plotted in Fig. 12.15, material velocity plotted in Fig. 12.16, and welding pressure plotted in Fig. 12.17 are consistent in each of the 27 design of experiments and is in good agreement with the results obtained by Donati et al. [17] and Liu et al. [12]. Based on similar results in their research the characteristics of effective stress, material velocity and weld pressure can be summarized by sectioning the welding surface into three regions:

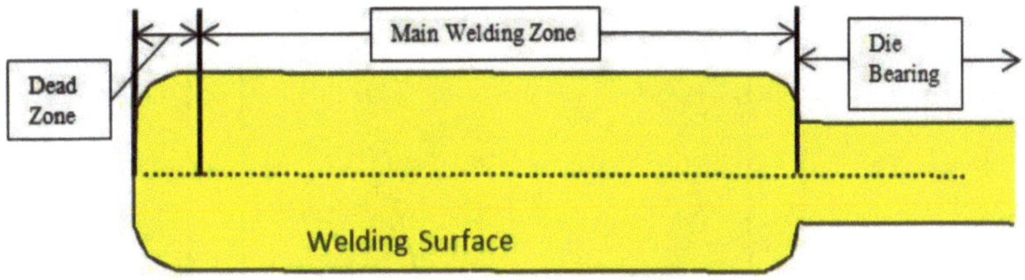

Fig. 12.14 Welding surface sections

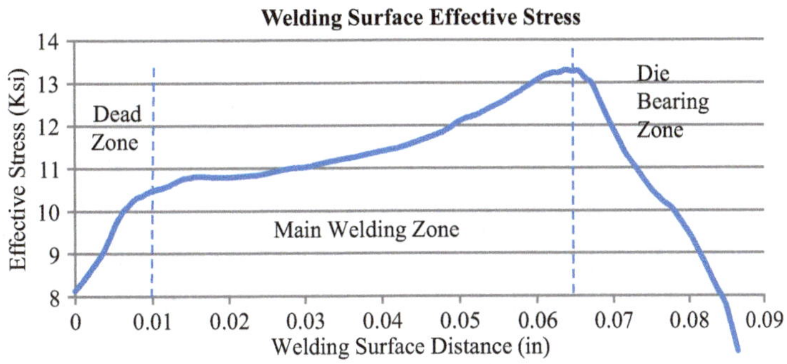

Fig. 12.15 Effective stress on welding surface

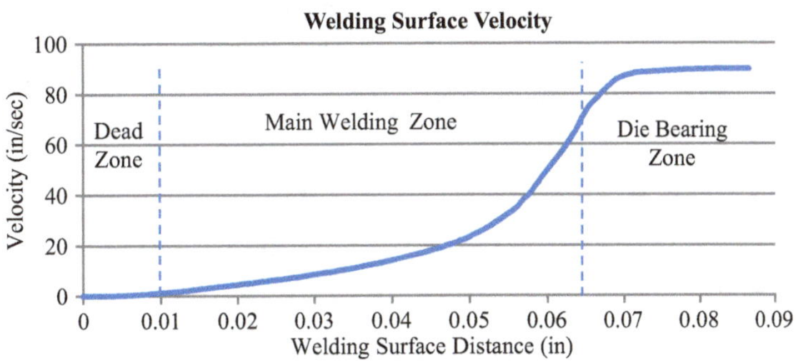

Fig. 12.16 Material velocity on welding surface

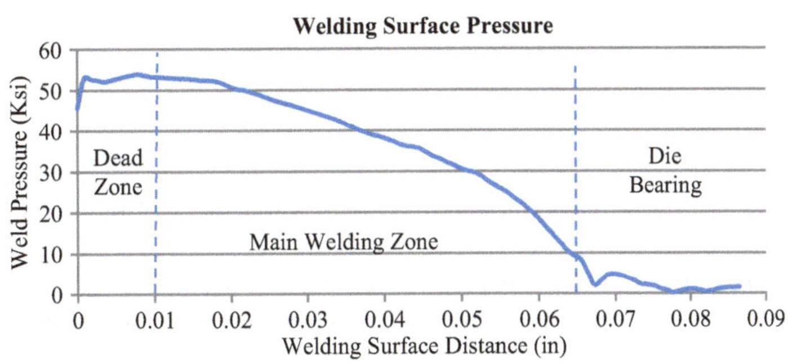

Fig. 12.17 Welding pressure distribution on welding surface

1. Dead Zone: is formed right under the entrance to the port opening and is characterized by very low velocity, maximum welding pressure and a gradual increase in effective stress from a relatively low value.
2. Main Welding zone: is formed below the dead zone and is characterized by increase in material velocity while welding pressure decreases and effective stress gradually increases to a maximum value.
3. Die Bearing Zone: is the final region where the final shape of the MMP tube is formed and is characterized by highest velocity while the welding pressure is the lowest and effective stress gradually decreases from its maximum value at the beginning of the die bearing zone due to the maximum deformation state of the extrusion profile.

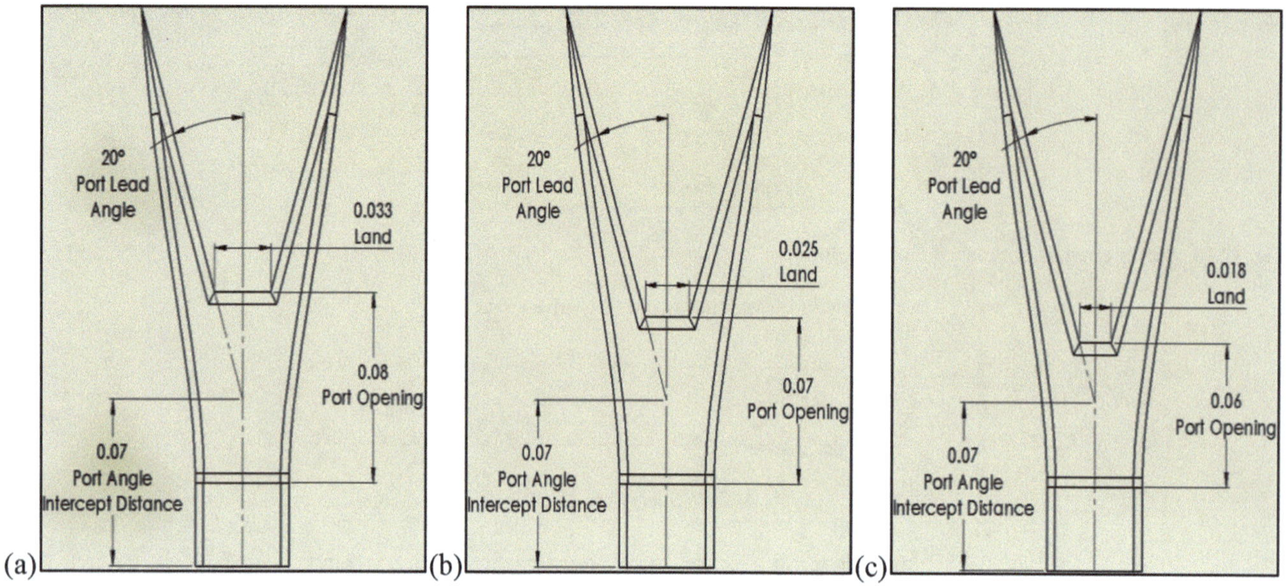

Fig. 12.18 Port opening; (**a**) 0.06inch; (**b**) 0.07inch & (**c**) 0.08inch

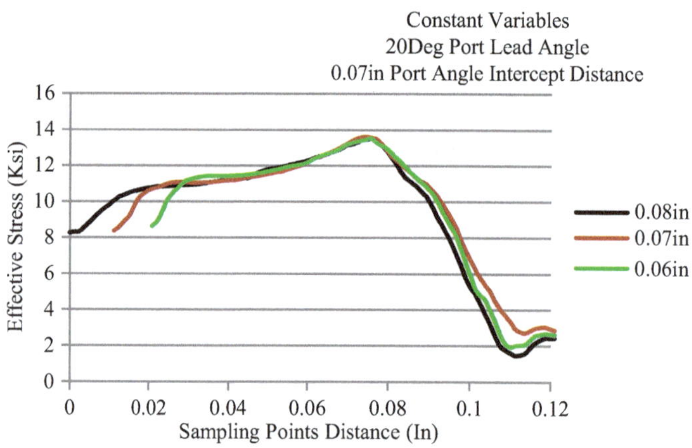

Fig. 12.19 Effective stress Vs. port opening on welding surface

12.3.2.2 Effect of Port Opening

The purpose of the port opening is to allow the aluminum to weld back together before the thin walls of the condenser tube are formed.

Figure 12.18 shows the CAD geometry of the mandrel with varying port opening lengths. It is observed that the length of the land is 0.018 inches for 0.06 inch port opening length, 0.025 inches for 0.07 inch port opening length and 0.032 inches for 0.08 inch port opening length.

Effective Stress

Figure 12.19 compares the effective stress of the deforming aluminum on the welding surface by varying port opening lengths. It is observed that trend of effective stress in the dead zone, main welding zone and die bearing zone is similar in all port opening lengths. However, the length of the main welding zone decreases due to the decrease in port opening length.

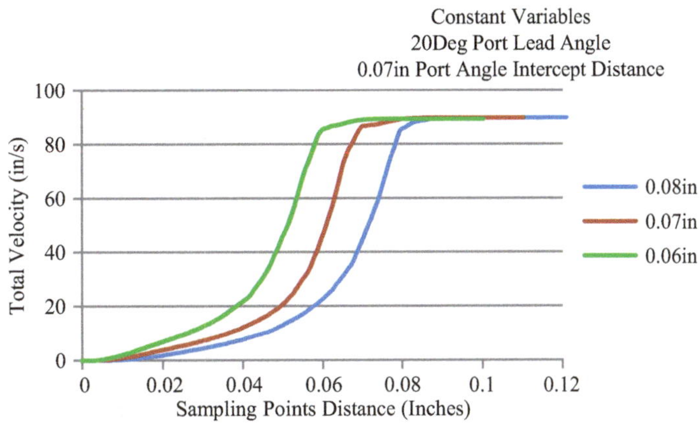

Fig. 12.20 Velocity vs. port opening on welding surface

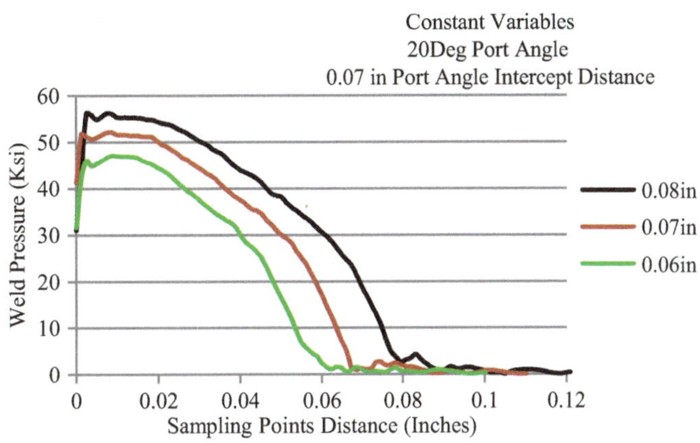

Fig. 12.21 Weld pressure Vs. port opening on welding surface

Material Velocity

Figure 12.20 compares the flow of aluminum on the welding surface by varying port opening lengths. It is observed that the velocity of aluminum increases relatively faster when the length of the port opening decreases. Also, the dead zone of aluminum increases as the port opening length increases, this behavior of aluminum flow is mainly dictated by the increase in length of land at the port opening entrance which increases as the length of the port opening increase. However, the exit velocity of aluminum in all varying cases is similar due to the constant extrusion ratio.

$$Extrusion\ Ratio\ (ER) = \frac{Cross\ sectiona\ area\ of\ the\ billet\ (A_b)}{Cross\ sectiona\ l\ area\ of\ the\ profile\ \left(A_p\right)} \tag{12.5}$$

$$Exit\ Velocity\ \ (V_e) = Ram\ Velocity\ (V_r) \times Extrusion\ Ratio\ (ER) \tag{12.6}$$

Welding Pressure

Figure 12.21 compares the weld pressure distribution of varying port opening lengths. It is observed that the welding pressure increases as the length of the port opening increases. This trend of weld pressure distribution is characterized by the combined effect of longer main welding zone observed in the effective stress distribution and longer dead zones observed in the material velocity distribution (Fig. 12.20) of varying port lengths. Therefore, 0.08 inch port opening length has the highest welding pressure distribution whereas, 0.06 inch port opening length has the lowest weld pressure distribution on the welding surface.

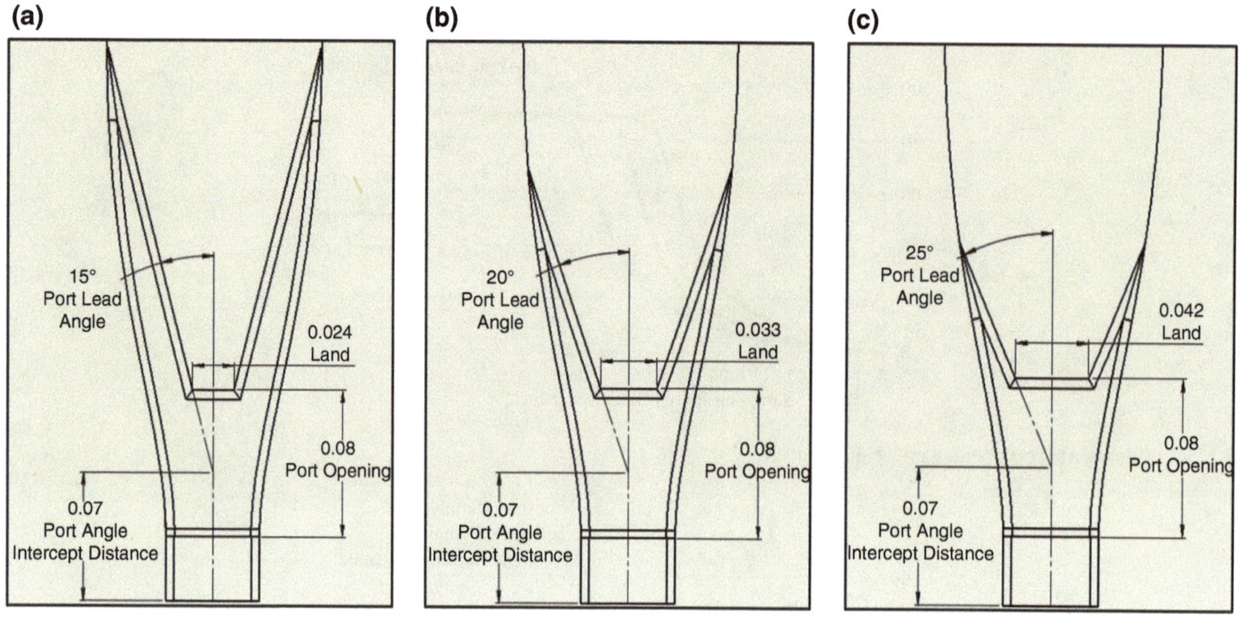

Fig. 12.22 Port lead angles; (**a**) 15°; (**b**) 20°, (**c**) 25°

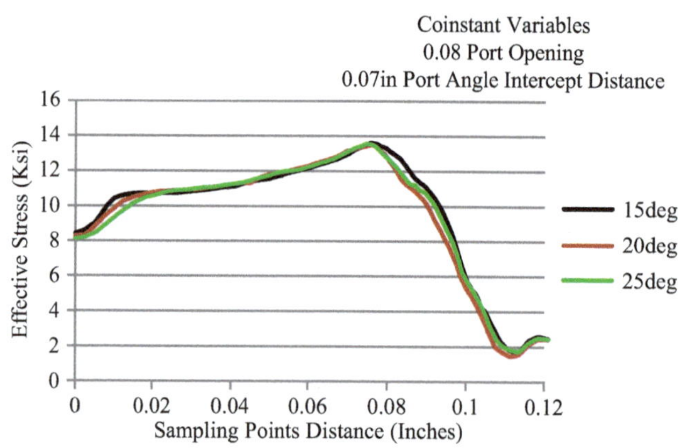

Fig. 12.23 Effective stress vs. port lead angle on welding surface

12.3.2.3 Effect of Port Lead Angle

The purpose of the port lead angle is to guide the aluminum flow from the porthole chamber into the port opening in order to gaining maximum weld. Figure 12.22 shows the CAD geometry of the mandrel with varying port lead angles. It is observed that increasing the port lead angle restricts the total volume of the porting.

Effective Stress

Figure 12.23 compares the effective stress of the deforming aluminum on the welding surface by varying port lead angles. It is observed that the effective stress in the dead zone decreases with increasing port lead angle which gradually becomes constant over the main welding zone and then decreases in the die bearing zone as the port lead angle increases.

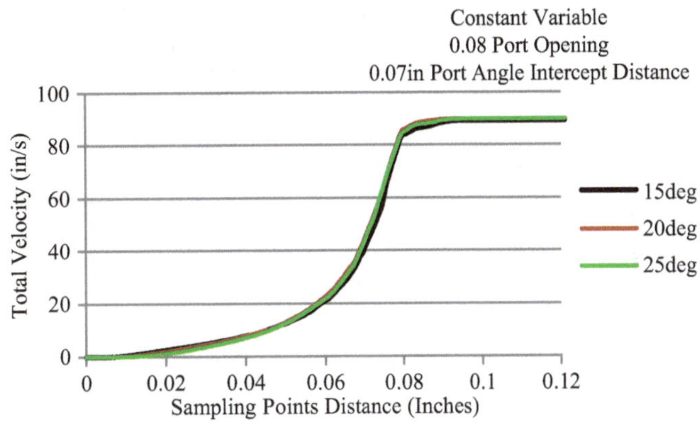

Fig. 12.24 Velocity Vs. port lead angle on welding surface

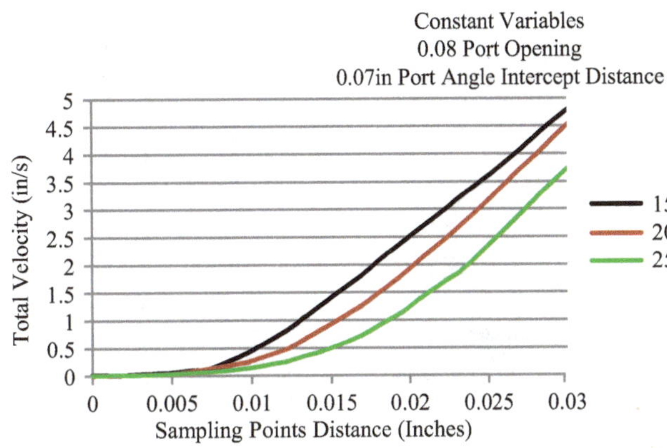

Fig. 12.25 Velocity in dead zone Vs. port lead angle on welding surface

Material Velocity

Figure 12.24 compares the flow of aluminum on the welding surface by varying port lead angle. It is observed that the velocity profile over the entire welding surface follows a similar trend in all varying port lead angles. However, Fig. 12.25 shows the dead zone increases with increasing angle due to the increase in length of the flat land.

Welding Pressure

Figure 12.26 compares the weld pressure with the varying port lead angles on the welding surface. It is observed that the changes in weld pressure pertaining to the port lead angle is very minute, where the welding pressure for 25° port lead angle is 53.75 Ksi, 20° port lead angle is approximately 55.14 Ksi and 15° port lead angle is approximately 57.24 Ksi. This trend of weld pressure observed at the entrance of the port opening on the welding surface is characterized by high effective stress and small dead zone of aluminum flow due to the increase of the port lead angle.

12.3.2.4 Effect of Angle Intercept Distance

The purpose of the port angle intercept distance is to hypothetically mark the point at which welding will commence in the port opening. This gives the die designer a rough idea of how big the dead metal zone will be in the port opening thus predicting weld formations.

Figure 12.27 shows the CAD geometry of the mandrel with varying angle intercept distance. It is observed that the length of the land increases that is 0.055 inches for 0.04 inch angle intercept distance, 0.033 inches for 0.07 inch angle intercept distance and 0.011 inches for 0.1 inch angle intercept distance.

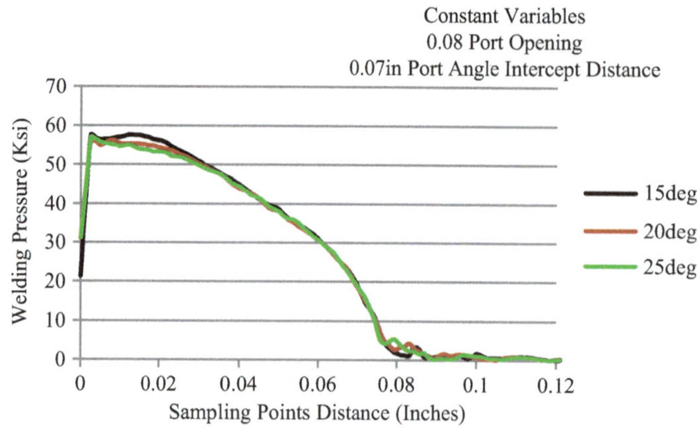

Fig. 12.26 Weld pressure Vs. port lead angle on welding surface

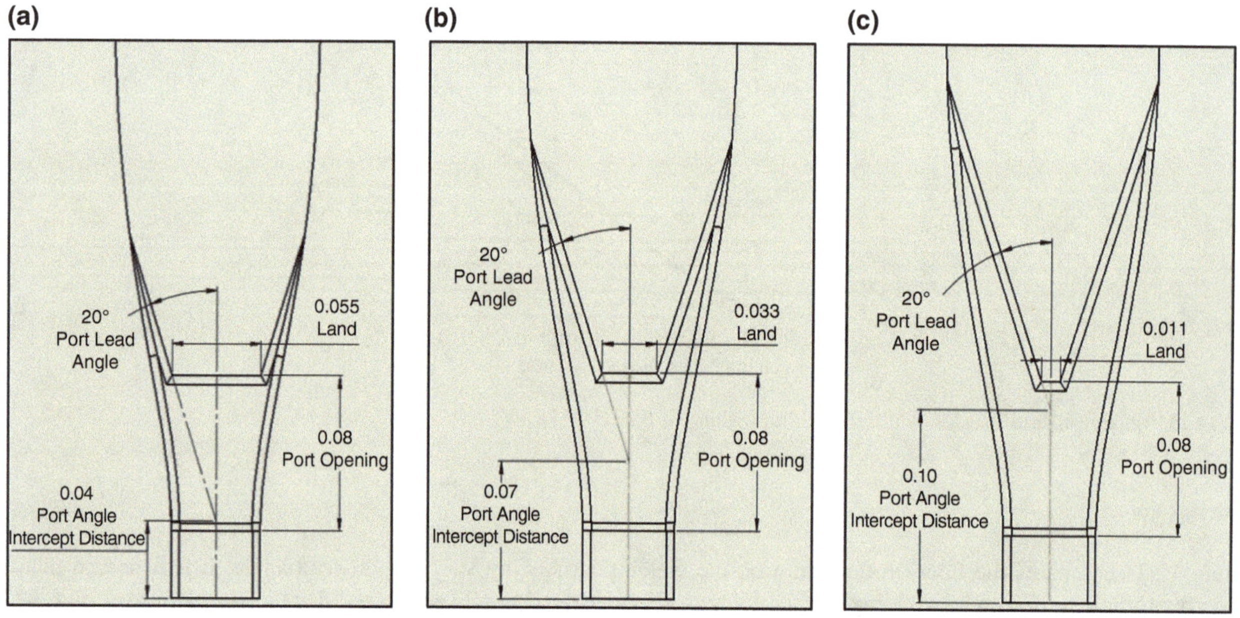

Fig. 12.27 Angle intercept distance; (**a**) 0.04inch; (**b**) 0.07inch & (**c**) 0.1inch

Effective Stress

Figure 12.28 compares the effective stress of aluminum by varying the angle intercept distance. The first peak in the effective stress graph marks the onset of the main welding zone. It is observed that the length of the dead zone decreases as the angle intercept distance increases, whereas the length of the die bearing zone decreases.

Material Velocity

Figure 12.29 compares the velocity of aluminum by varying the angle intercept distance. It is observed that angle intercept distance does not significantly affect the aluminum velocity distribution. The 0.1inch angle intercept distance has the smallest land length; therefore, when comparing the aluminum flow in Fig. 12.30, it is observed that the dead metal zone is the smallest for 0.1inch angle intercept distance and is the biggest for 0.04 inch angle intercept distance.

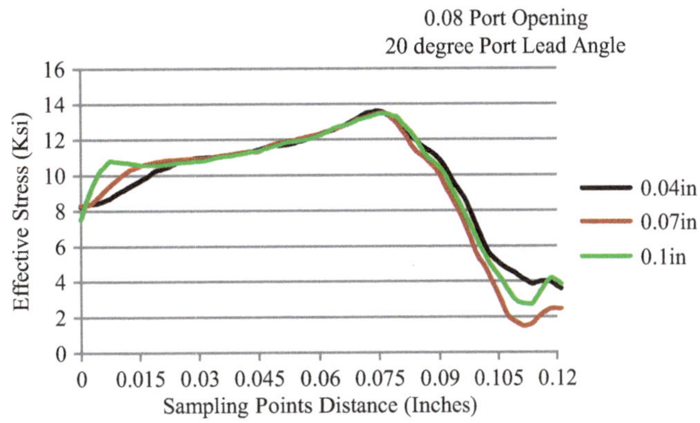

Fig. 12.28 Effective stress Vs. angle intercept distance on welding surface

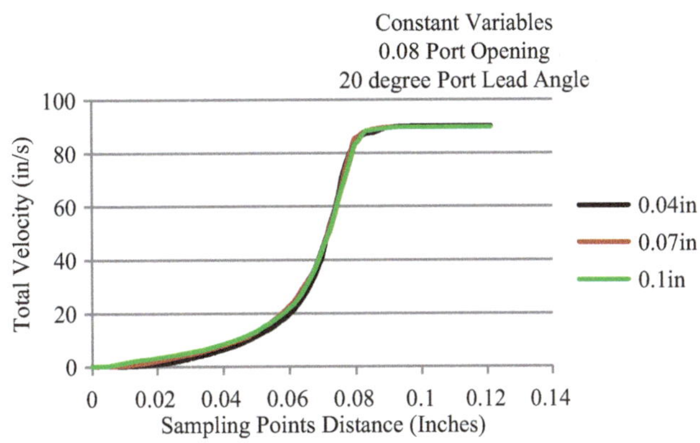

Fig. 12.29 Velocity Vs. angle intercept distance on welding surface

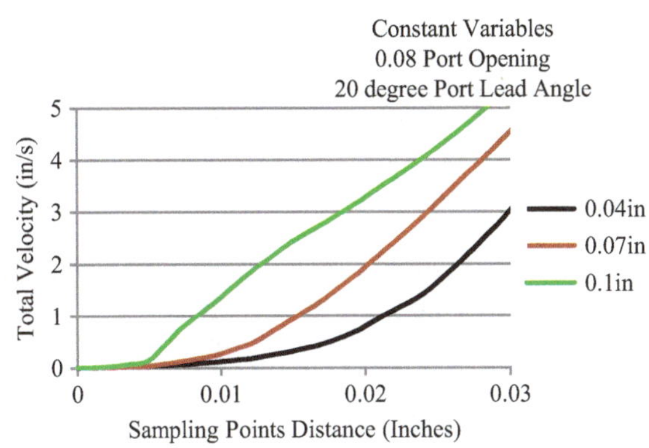

Fig. 12.30 Velocity in dead zone Vs. angle intercept distance on welding surface

Welding Pressure

Figure 12.31 compares the weld pressure on the welding surface with varying angle intercept distance. The 0.1 inch angle intercept distance has a higher welding pressure of 60.56Ksi due to the combined effect of relatively high effective stress at the entrance of the port opening on the welding surface and a very small dead zone observed in the aluminum flow. The welding pressure observed for 0.07 inch angle intercept is 55.29Ksi and 0.04 inch angle intercept distance is 53.48Ksi. The maximum welding pressure achieved by the 0.1 inch angle intercept distance is due to the smallest length of land, thus leading to the formation of a very small dead zone and a relatively long main welding zone.

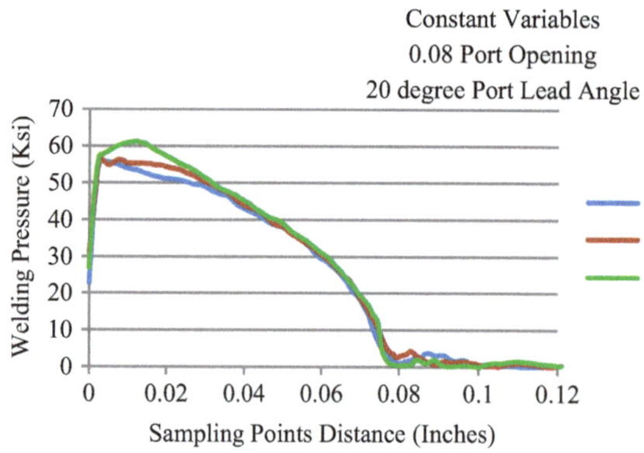

Fig. 12.31 Weld pressure Vs. angle intercept distance on welding surface

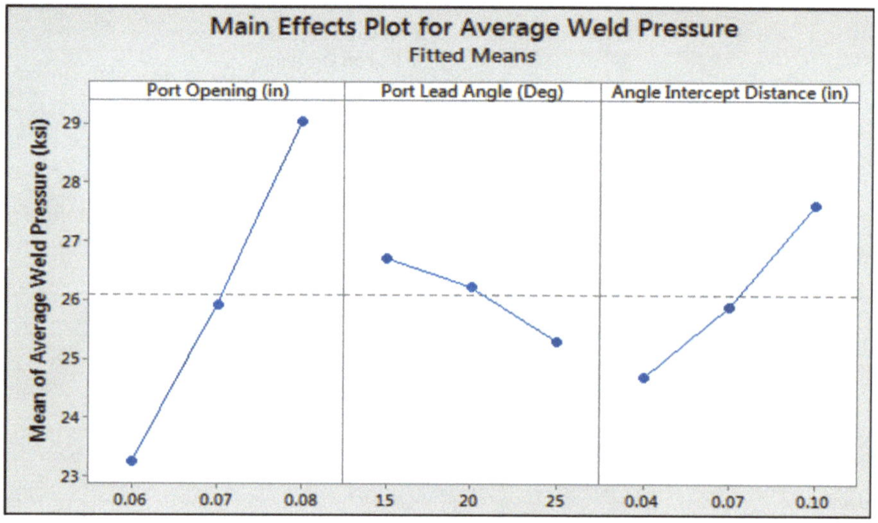

Fig. 12.32 Main effects plot for average weld pressure

12.3.2.5 Statistical Summary

A full factorial design of experiments was conducted and statistically analyzed using Minitab, to investigate the effects of mandrel port opening length, port lead angle and angle intercept distance on weld pressure. A main effects plot is an efficient visual tool in Minitab that summarizes the effect of each independent geometric parameter examined in all 27 mandrel DOE's.

Figure 12.32 shows the main effects plot generated by analyzing the mean values of weld pressure at each level of varying port opening length, port lead angle and angle intercept distance. For example the mean weld pressure is 23.27 ksi for all nine possible combinations of mandrel DOE with 0.06 inch port opening length The average value of weld pressure for all 27 mandrel DOE's is 26.09 ksi, which is represented as a horizontal dashed line shown in the main effects plot. The left and right panel in the main effects plot indicates that increasing the length of the port opening and the angle intercept distance increases the weld pressure. However, the middle panel in the main effects plot indicates an inverse effect on the weld pressure which decreases as the port lead angle increases.

Although, the information provided by the main effects plot help guide the intuition of the die designer it is important to know which ones out of the three mandrel geometric parameters have a significant effect on welding pressure. To understand which mandrel geometric factors significantly affects the response of the weld pressure we compare the probability values also known as *P*-value generated through the full factorial analysis in Minitab. The definition of a P-value is the probability of either accepting or rejecting a null hypothesis based on the threshold of a confidence interval which is typically 95% or a significance interval (α) of 0.05.

$$\mathit{Confidence\ Interval} = 100\,(1 - \alpha)\,\% \tag{12.7}$$

Source	P-Value
Model	0.000
Linear	0.000
Port Opening (in)	0.000
Port Lead Angle (Deg)	0.000
Angle Intercept Distance (in)	0.000
2-Way Interactions	0.001
Port Opening (in)*Port Lead Angle (Deg)	0.652
Port Opening (in)*Angle Intercept Distance (in)	0.000
Port Lead Angle (Deg)*Angle Intercept Distance (in)	0.012

Fig. 12.33 Minitab general factorial regression model *P* value

Figure 12.33 shows the p values of all independent factors and their two-way interactions. It is observed that port opening length, port lead angle and angle intercept distance are statistically significant factors as their P-Values are less than 0.05 as well as, the interaction of port opening and angle intercept distance. This means that changing any of these parameters has a significant effect on weld pressure.

12.4 Conclusion

The main objective of the work presented, was to identify and use a suitable extrusion simulation package that can assist die designers predict the effect of changing a certain geometric parameter in the die design on the aluminum flow and deformation of the die tooling components. The main contributions of this work can be summarized as follows:

1. A methodological design framework using a commercial extrusion simulation package was established for the analysis of die stress distribution and aluminum flow through the modular die for the production of AA3003 aluminum alloy Micro-Multiport tubes used for manufacturing of contemporary heat exchange applications.
2. Using the Archards wear model; based on abrasive wear theory. Reliable predictions on the mechanics of wear as well as, amount of wear depth, temperature distribution and stress concentration of each coating layer in the mandrel and die plate were analyzed. It was predicted that the service life of a bilayer coated mandrel is 12.07% higher than the non-coated mandrel, whereas the service life of the coated die plate is 6.63% higher than the non-coated die plate.
3. A design of experiments was conducted to understand the effects and interactions of port opening, port lead angle and port angle intercept distance; of the mandrel CAD geometry on weld pressure, velocity and effective stress of aluminum on the welding surface in the MMP tubes. Using Minitab a statistical analysis software, it was found that port opening and port angle intercept distance have a significant effect on the over quality of weld pressure.
4. A design of experiments was also conducted to understand the effects and interactions of angle guide curve, die plate covering and bridge mandrel angle; of the die cartridge CAD geometry leading to the deformation stresses of the mandrel teeth. Through the statistical analysis conducted for this DOE, it was found that die plate covering has a significant effect on the deformation of the mandrel teeth.
5. The Minitab response optimizer was utilized for the optimization of the geometric parameters for the mandrel and die cartridge. Which was validated by comparing the results with the Kriging optimization that provide the global optimized results.

    (a) The optimum combination for the mandrel geometric parameters that resulted in a maximum weld pressure of 42.45Ksi is 0.08 inch port opening, 20° port lead angle and 0.1 inch angle intercept distance.
    (b) The optimum combination for the die cartridge geometric parameters that resulted in a minimum deformation of the mandrel teeth is 15° angle guide curve, 0.9 inch die plate covering and 20° bridge mandrel angle. The results of the die cartridge design of experiments ignore the effects of wear on the die plate. Hence 0.9 inch die plate covering is feasible.

References

1. He, Y.-F., et al.: FEM simulation of aluminum extrusion process in porthole die with pockets. Trans. Nonferrous Metals Soc. China. **20**(6), 1067–1071 (2010)
2. Oh, S.I., Wu, W.T., Tang, J.P., Vedhanayagam, A.: Capabilities and applications of FEM code DEFORM: the prespective of the developer. J. Mater. Process. Technol. **27**, 25–42 (1991)
3. Li, G., et al.: Recent development and applications of three-dimensional finite element modeling in bulk forming processes. J. Mater. Process. Technol. **113**(1–3), 40–45 (2001)
4. Wang, L., Yang, H.: Friction in aluminium extrusion—part 2: a review of friction models for aluminium extrusion. Tribol. Int. **56**(0), 99–106 (2012)
5. Sheppard, I.F.T.: Nature of friction in extrusion process and its effect on material flow. Mater. Sci. Technol. **19**(7), 837–846 (2003)
6. Wang, L.: Modelling of Friction for High Temperature Extrusion of Aluminium Alloys, in Engineering. Harbin Institute of Technology (2012)
7. Tushar Bakhtiani, H.E.-M., Zhang, J.: Modeling of extrusion process of a condenser tube for investigating the effects of mandrel geometry. Int. J. Adv. Manuf. Technol. 1–16 (2016)
8. Sofuoglu, H., Gedikli, H.: Physical and numerical analysis of three dimensional extrusion process. Comput. Mater. Sci. **31**(1–2), 113–124 (2004)
9. Arentoft, M., et al.: Physical and mathematical modelling of extrusion processes. J. Mater. Process. Technol. **106**(1–3), 2–7 (2000)
10. Xianghong, W., et al.: Numerical simulation and die structure optimization of an aluminum rectangular hollow pipe extrusion process. Mater. Sci. Eng. A. **435–436**(0), 266–274 (2006)
11. Khansai, Y.A.-U.-L.K.: 3D numerical simulation to investigate the influential factors causing die failure in hot Aluminum extrusion of square hollow profile. Int. J. Sci. Eng. **2**(2), 11–17 (2012)
12. Liu, J., et al.: Effects of process parameters and die geometry on longitudinal welds quality in aluminum porthole die extrusion process. J. Cent. S. Univ. Technol. **17**(4), 688–696 (2010)
13. Saha, P.K.: Thermodynamics and tribology in aluminum extrusion. Wear. **218**(2), 179–190 (1998)
14. H.G. Mooi, A.J. Den Bakker, K.E. Nilsen, J. Huetink, Simulation of Aluminum Extrusion based on a Finite Element Method (FEM). (1991)
15. Tomesani, L.D.L.: Analysis of material flow and welding in aluminum extrusion of hollow sections. In: V AITEM Conference, 2001: pp. 678–698
16. Flitta, I., Sheppard, T.: Material Flow and Prediction of Extrusion Pressure When Extruding Through Bridge Dies Using FEM (2000)
17. Donati, L., Tomesani, L.: The prediction of seam welds quality in aluminum extrusion. J. Mater. Process. Technol. **153–154**(0), 366–373 (2004)